Circularity Assessment: Macro to Nano

Rashmi Anoop Patil ·
Seeram Ramakrishna
Editors

Circularity Assessment: Macro to Nano

Accountability Towards
Sustainability

 Springer

Editors
Rashmi Anoop Patil
National University of Singapore
Singapore, Singapore

Seeram Ramakrishna
National University of Singapore
Singapore, Singapore

This work was supported by Olympia Electronics SA

ISBN 978-981-19-9702-0   ISBN 978-981-19-9700-6   (eBook)
https://doi.org/10.1007/978-981-19-9700-6

This Springer imprint is published by the registered company Springer Nature Singapore Pte Ltd.
The registered company address is: 152 Beach Road, #21-01/04 Gateway East, Singapore
189721, Singapore

To sustainable and resilient living that I wish becomes the new global conscience soon...

—Rashmi Anoop Patil

Foreword

We are witnessing the greatest climate risk of our times which is disrupting the economic and investment market, and concurrently with the geo-political uncertainties, exposing the supply chain and energy system vulnerabilities on a global scale. This calls for a pragmatic transformation from the current linear economic model to a more resilient and sustainable circular one where there are reduced dependencies on natural resources and minimized risks due to a diversified supply chain. In recent years, the World Bank, the European Union, the World Economic Forum, Cloud Computing global providers together with leading Enterprise and consulting organizations are struggling to maintain business as usual and are using "the Clean/Green Earth vision toward sustainability for future generations" to compete for differentiation and maintain their growth pace. The World Bank has emphasized the importance of transitioning to a circular economy and having an integrated approach to circularity assessment, which includes the conscious participation and cooperation of governments, businesses, and communities. Recent research reports on the trends in circular economy and circularity assessment collectively highlight key avenues that can facilitate a successful transition to a circular economy, such as (i) systemic changes (in business practices, consumer behavior, and government policies), (ii) significant investment (in infrastructure, technology, and innovation), and (iii) creation of new business opportunities (based on products and services that use waste as a resource). In agreement with this, many countries, local administrations, global businesses, and start-ups across the world are embracing circular economy as a means to achieve sustainability and consequently engaging themselves in lowering the carbon footprint towards fulfilling the objectives of the Paris Agreement.

This brings us to the question- how is the progress toward circularity being evaluated? Currently, a lack of expertise in quantifying the pursuit of a circular economy and the absence of standardized evaluation guidelines are hampering our understanding of the status quo. Varied methods of assessment using different sets of indicators, less-reliable data, unaccounted interdependencies, and exclusion of feedback to the stakeholders are the major factors contributing to poor evaluations. It's clear from this that there is a need for a standardized evaluation and feedback system to provide momentum toward sustainability while capitalizing on the potential benefits presented by the transition to a low-carbon circular economy.

The authors of this work have articulated the essentiality of introducing a standardized circularity evaluation system at various levels of economic complexity by bringing out the inefficiencies of state-of-the-art evaluation systems. Such a comprehensive evaluation system has the potential to lead the stakeholders in the right direction in pursuit of a circular/sustainable future by providing feedback on their progress (closing the loop). The importance of such an evaluation system is highlighted by presenting the areas for improvement in state-of-the-art circularity assessment methodologies and the advantages of bringing them together for a comprehensive analysis and understanding of circularity status. Lastly, and more importantly, the authors emphasize how conscious consumerism has a critical and constructive role to play in this profound transition.

Overall futuristic roadmap for the development and integration of circularity assessment at various systemic hierarchies should include the development of a common conscience of circularity and tools that help all the stakeholders effectively assess and improve the circularity of every part of the economic system. Any circularity assessment exercise should take into consideration key aspects such as (i) robust metrics and indicators, (ii) comprehensive assessment tools, (iii) a more effective engagement with stakeholders inclusive of building new collaborative networks, (iv) the development of an extensive database of indicators, inclusive of real-time updation and continuous management, (v) development of new policies and regulations and effective implementation of the same, and (vi) raising awareness among the stakeholder and educating the consumers regarding their role in the pursuit of circularity.

Any effort to build a comprehensive closed-loop circularity assessment framework that can be universally applied to various economic scenarios faces a catch-22 situation - in the form of a disconnect between the underlying intent coupled with the ongoing efforts towards circularity and the requirements of a transparent roadmap for achieving comprehensive circularity in a timely manner. In the long run, a well-understood and realized closed-loop circularity evaluation system can be beneficial, outweighing the costs and the time required to reach there. This may eventually assist in flushing out green-washing in businesses and ensure that the pursuit towards circularity progresses in the intended direction, largely helped by periodic assessments and milestone-based outlooks.

I found this book a must read, it is the pillar of a transformation change and will establish the opportunity at stake for academicians, researchers, and most importantly, students, to appreciate the need for a basic yet thorough understanding of the current assessment efforts and the inherent gaps to be addressed critically. I appreciate the authors for their contributions to this work and the efforts of the editors- Ms. Rashmi Anoop Patil & Prof. Seeram Ramakrishna, in putting together this book. Furthermore, I hope this report can serve as a core foundation and a motivation for further studies, research, and action in the form of policies and business directions spanning developed and developing countries across the world, and drive a progressive economic and social conscience toward a circular society.

Stanford, CA, USA
March 2023

Susanna Kass
Co-founder, BOD, InfraPrime;
Global Management Board (GMB);
Stanford University Graduate
School of Business (GSB) Women's Circle
Leading Impact, Analysis & Reporting
Topmost 50 Climate Change Leader;
Data Center Advisor, UNSDG-EP;
Top 10 Most Inspiring Leaders
to follow in 2023, Inc. Magazine;
Top 10 Data Center Influencers
to follow on LinkedIn;
Top 50 Women Leaders in San Francisco
in 2022 by Women Most Admired.

Preface

The primary motivation for this book is to present the state-of-the-art, widely implemented methods and indicators for circularity assessment at different levels of an economic system inclusive of urban areas/large geographical regions, a business sector, individual businesses, and products and services. The book provides a general perspective on the circularity assessment at different levels of the systemic hierarchy and advocates better resource management, environmental and social performances for a sustainable future. This work caters to a broad readership across various fields inclusive of governance, research, academics, and businesses.

The book starts with the foundational concepts of circularity assessment in Chap. 1. The chapter focuses on the importance and the need for circularity assessment as an enabler for the transition from a linear to a circular economy (CE). This is followed by the conceptual perspective of the various systemic levels that can be considered for the circularity assessment process. A brief discussion on the need for a standard and universally applicable framework for circularity assessment then follows. Such a framework needs (i) a comprehensive set of indicators that can quantify the circularity extent and/or performance of a particular systemic level and (ii) an ISO standard for circularity assessment, in times of growing relevance and conscience of circularity around the world. The authors, thus, emphasize the need for accountability towards a circular future, in which the performance assessment of a systemic level (regarding how circular a systemic level is) provides the insights and direction for the transition to CE.

In recent times, the development of various global or stock exchange-related standards and government regulations for sustainability reporting, popularly known as Environmental, Social, and Governance (ESG) reporting, has led to an increase in the number of companies evaluating and reporting their performance across the globe. ESG reporting also includes many metrics related to circularity in both environmental and social aspects. This process of framing standards and regulations provides insights for developing an evaluatory framework for circularity assessment. The trends in ESG reporting and the adoption of a sustainability mindset within the investor community have majorly influenced the financial sector to introduce sustainable finance for development projects towards sustainability (and implicitly, circularity). To tap this potential for financing new circular projects, there is a need to incorporate circularity assessment within ESG reporting. Chapter 2 explores what lessons can be learnt from the evolving and iterative process of developing ESG standards and highlights how the inclusion of circularity assessment within sustainability reporting helps in raising funds for circular projects as well.

Chapters 3–6 discuss the circularity assessment at the four systemic levels, starting from the macro-level—urban areas/countries, followed by eco-industrial parks/industrial sectors (at the meso-level), individual businesses (at the micro-level), and individual products/services (at the nano-level). At each of these levels, we try to (i) present a comprehensive set of relevant indicators for quantifying the circularity, (ii) a discussion on the qualitative and quantitative interpretation of the progress towards circularity, and (iii) develop know-how on how each of those levels functions to render themselves circular in their operations and management through case studies. For the macro-level, we consider Rotterdam and Paris as representative examples for urban areas that are embracing circularity with decent success. For the meso-level, we study the eco-industrial parks in Austria, South Korea (Ulsan), and Denmark (Kalundborg), and then Apple Inc. as a representative case study for businesses that have made significant strides in embracing circularity in their operations. This is followed by a discussion on how circularity can be embraced and measured at the nano-level. Case studies on Levi's jeans and single-use surgical face masks serve as examples for analyzing the environmental impact of a product's life cycle using relevant indicators.

In the concluding chapter, we present the concept of granular circularity centered around an individual consumer. Realizing the vision for consumer-centric circularity is dependant on (i) the ripple-effect caused by the efforts of a self-motivated individual/group of consumers leading to a significant change in the society towards circularity, (ii) the role of governments in promoting and embracing circularity, (iii) understanding the benefits and challenges of a circular lifestyle. Closing the circularity gap is finally presented as a part of the authors' opinion and a logical end to the content presented in this book.

Singapore   Rashmi Anoop Patil
                                                              Patrizia Ghisellini
                                                         Sven Kevin van Langen
                                                          Seeram Ramakrishna

Acknowledgements

We are deeply grateful to all the authors of this book for their invaluable contribution to this work.

This work has received funding for open-access publishing from

i. the European Union's Horizon 2020 Research and Innovation programme under the Marie Skłodowska-Curie Innovative Training Networks (H2020-MSCAITN-2018) scheme, grant agreement number 814247 (ReTraCE project).

The authors Dr. Patrizia Ghisellini and Sven Kevin van Langen have received funding for their research projects from

i. the European Commission's research programme Horizon 2020-SC5-2020-2 scheme, Grant Agreement 101003491 (JUST Transition to the Circular Economy project),
ii. the China-Italy High Relevance Bilateral Project funded by the Ministry of Foreign Affairs and International Cooperation (MAECI), the General Directorate for the Promotion of the Country System (Grant No. PGR00954),
iii. the European Union Project EUFORIE, European Futures for Energy Efficiency, funded under H2020-EU.3.3.6.—Robust decision making and public engagement, and
iv. the European Union's Horizon 2020 research and innovation programme under the Marie Skłodowska-Curie Innovative Training Networks (H2020-MSCA-ITN-2018) scheme, grant agreement number 814247 (ReTraCE project).

The original sketch of the cover page artwork was provided by Ms. Rashmi Anoop Patil. We express our special thanks to Dr. Anoop C. Patil (Wilmar Intl, SG) for the cover page design, and for his inputs and excellent support throughout the development of this book.

June 2022

Rashmi Anoop Patil
Seeram Ramakrishna

Prologue

Circularity, Circular Society, Circular Economy

Circularity is older than mankind and has gone through four distinct phases.
First was **Nature's** circularity by evolution, with natural cycles such as
marine tides, CO_2, and water cycles, plants, and animals. There is no waste;
dead material becomes food for other animals or plants. And Nature has no
preferences. Early mankind's **survival** depended on frugal and skilled use
of local resources. People and nature shared a non-monetary and chaotic
symbiosis dominated by Nature. The motivation for the circularity of early
Societies came from **necessity** or scarcity.

Phase 2—the **Anthropocene** started in **1945** and created mass-produced
'synthetic' man-made materials and energy. These were superior to natural
resources and gave mankind independence from Nature.

Phase 3—**Today, invisible quality** cycles of the immaterial world, such as
cultural values, sharing, caring, innovation, and responsibility, civil society
now becomes a key player.

Phase 4—The Future: Nature and man will live in synergy—or mankind
may notsurvive.

Phase 1—Circularity is built on managing the **use of assets**—a synonym
for stocks or capitals—of natural, human, cultural, manufactured, financial
or immaterial character. Embodiedwater and CO_2 emissions are examples of
immaterial assets.

On a **personal** level, Circular Society means to enjoy the use of one's
belongings and take good care of them, based on personal values and a
sense of sobriety.

On a **societal** level, Circular Society means to take good care of all assets
and optimize their long-term use, based on regional **cultural** values, and
live from the dividends of these assets.

Two renewable assets need our special attention:

- **Water:** because no resource can replace it, and because cleanwater is a
  necessity for the health and survival of people and animals.
- **Labour:** because people are the only resource with a qualitative edge,
  which can be greatly improved through education and training, but will
  rapidly degrade if unused.

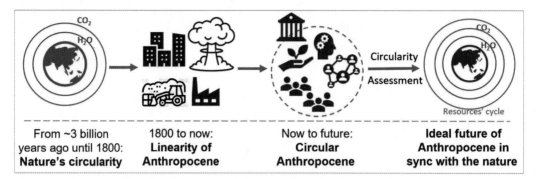

Fig. 1 A schematic representation of the four phases in the ongoing journey of circularity. The leftmost part (Phase 1) shows nature's circularity by evolution and the next stage (Phase 2) represents human activities adversely impacting the environment resulting in a linear economy. Currently, we find ourselves (Phase 3—third from left) attempting to charter towards a circular economy (Phase 4–symbolized in the rightmost part), through innovations, responsible behavior, and valuecontributions by the stakeholders (such as producers, manufacturers, businesses, consumers, and recyclers represented in the schematic illustration of phase 3), to be in synergy with nature. Most importantly, this process needs to be facilitated with relevant legislation and a holistic, regular assessment of the progress toward circularity.

Phase 2—The watershed of the Anthropocene started on 6 August 1945 in Hiroshima. When I was born 10 months after Hiroshima, few synthetic or man-made materials existed. There were no plastic or synthetic chemicals in the environment, not a single man-made object in Space, no computers or mobile phones, and few metal alloys (like brass). The world population was less than 2.5 billion people. 30 percent of people lived in cities, the biggest beingNewYork City, Tokyo, and London.

The Anthropocene helped to overcome scarcities of goods, shelter, and food but also created the need for a **Circular Industrial Economy** to take care of derelict objects and man-made materials incompatible with Nature's circularity. Note that a Circular Industrial Economy develops through a conscious decision by economic actors, not by business as usual.

Tackling the synthetic waste **legacy** of the Anthropocene dictates to close **three** loops:

- **Loop 1 to maintain the utility and value** of infrastructure, buildings, and objects by extending their service life—the era of R, which is local and intensive in skilled labour. In the era of R, owner-users are in charge of their belongings—you and me, economic actors as well as public authorities.
- **Loop 2 to maintain the purity and value of material assets**, by recoveringatoms and molecules—the era of D, which is regional and capital intensive. Theera of D suffers from a lack of responsibility on who is to collect, sort, andseparate used materials, and finally to de-link composite materials.
- **Loop 3 to maintain the liability of producers**, which are most qualified to take care of, and **pay** for the reuse of the derelict objects and used materials they created.

Phase 3—making use of the qualities of the immaterial world.
Only legislators can enforce closed liability loops.

Defining "waste" as manufactured objects and synthetic materials, without a positive value and an ultimate liable owner, opens two strategies

- To give "value" to derelict objects for instance by introducing deposit laws,
- To legislate ultimate liable owners such as the Polluter Pays Principle, and a mandatory take-back by producers—a reversed supply chain,

Smart manufacturers and fleet managers can pro-actively use such legislation to their advantage by shifting from the manufacturing economy to the Performance Economy

- By selling products as a service, through rental or operational leasing contracts, and
- by selling system performance (railways, airlines, chemical leasing, pay-peruse).

Retaining the ownership of objects means internalizing the liability for the costs of risk and waste, but also gives companies resilience, resource security, andcompetitive advantages.
"Today's objects are the resources of tomorrow, available locally at yesteryear's resource prices".
Retaining the ownership of objects and embodied materials also eliminates the compliance and transaction costs of the linear economy.
Politicians and public authorities can promote and profit from the Performance Economy by changing public procurement policy from buying objects to buying the service of objects, gaining in addition security for future supplies of resources.
A successful Circular Industrial Economy is built on economics, innovation, and competitiveness. Innovation is part of most university's curricula. The objective of Circular Economy Innovation is to develop circular sciences in the fields of energy, ceramics, chemistry, and metallurgy as well as a better understanding of behavioral and motivation sciences.
Phase 4—the Future Circular Society and the Circular Industrial Economy enable regions to turn today's major challenges into opportunities.
The transfer to a Circular Industrial Economy creates

- a **low waste** society through incentives to change individual behaviour—from consumer to user, and through loss and waste prevention by intelligent resource management.
- a **low carbon** society by preserving the water, electricity, and CO_2 emissions embodied in physical assets, and through innovation into green electricity and circular energy,
- a **low anthropogenic mass** society by preserving the existing stocks of infrastructure,buildings, equipment, vehicles, and objects, and develop circular material sciences.

So what is the circular economy?

- A regional economy that in the era of R replaces inputs of imported energy and virgin materials by the skilled labor of local workers optimising existing stocks.
- A disruptive economic model built on economics, innovation, and competitiveness.

Its substantial social, ecological, and economic benefits, compared to manufacturing, result from innovation into new materials, new business models, new technologies to reuse objects and recover used atoms and molecules—and an intelligent decentralization of the economy.

Ultimately it means exploring "The Limits to Certainty, facing risks in the New Service Economy".

References

1. Giarini, Orio, and Stahel, Walter R. (1993) The Limits to Certainty, facing risks in the new Service Economy, 2nd revised edition; Kluwer Academic Publishers, Dordrecht, Boston, London; ISBN 0-7923-2167-7. 270 p.
2. Stahel,Walter R (2010) The Performance Economy, 2nd edition. Palgrave Macmillan, Houndmills, ISBN 0-230-00796-1. 349 p. https://www.palgrave.com/de/book/9781349369195
3. Walter R. Stahel (2019) The Circular Economy—a user's guide. with a foreword by Dame Ellen MacArthur. Routledge, Abingdon https://www.routledge.com/The-Circular-Economy-A-Users-Guide-1st-Edition/Stahel/p/book/9780367200176

Geneva

24 February 2023

Walter R. Stahel

Founder-Director

The Product-Life Institute, Geneva;

Visiting Professor at the Faculty

of Engineering and Physical Sciences

University of Surrey;

Visiting Professor at Institut Environnement

Development Durable et Economie

Circulaire (IEDDEC), Montréal, Canada.

Contents

Contributors

Patrizia Ghisellini Department of Engineering, Parthenope University of Naples, Naples, Italy

Rashmi Anoop Patil The Circular Economy Task Force, National University of Singapore, Singapore, Singapore

Seeram Ramakrishna The Circular Economy Task Force, National University of Singapore, Singapore, Singapore

Sven Kevin van Langen UNESCO Chair in Environment, Resources and Sustainable Development (International Ph.D. Programme), Department of Science and Technology, Parthenope University of Naples, 80143 Naples, Italy;
Olympia Electronics, Thessaloniki, Greece

Acronyms

ATHEX	Athens Stock Exchange
CDP	Carbon Disclosure Project
CE	Circular Economy
CEIP	Circular Economy Indicator Prototype
CET	Circular Economy Toolkit
CMU	Circular Material Use
COVID-19	Coronavirus Disease of 2019
CSR	Corporate Social Responsibility
CSRC	China Securities Regulatory Commission
CTI	Circular Transition Indicators
EEA	European Economic Area
EFRAG	European Financial Reporting Advisory Group
EIP/s	Eco-Industrial Park/s
EMA	Emergy Accounting
EMEA	Europe, the Middle East, and Africa
ESG	Environmental, Social, and Governance
EU	European Union
EUROSTAT	Statistical Office of the European Communities
FSA	Financial Services Agency
GDP	Gross Domestic Product
GRI	Global Reporting Initiative
ICT	Information Communication Technology
IIRC	International Integrated Reporting Council
IPCC	Intergovernmental Panel on Climate Change
IR	Industrial Revolution
IS	Industrial Symbiosis
ISO	International Organization for Standardization
LCA	Life Cycle Assessment or Life Cycle Analysis
MCI	Material Circularity Indicator
MFA	Material Flow Accounting or Material Flow Analysis
NGO	Non-Governmental Organization
OECD	Organization for Economic Cooperation and Development
S&P	Standard and Poor
SASB	Sustainability Accounting Standards Board
SDG	Sustainable Development Goal
SEC	Securities and Exchange Commission

SLCA	Social Life Cycle Assessment
SMEs	Small and Medium Enterprises
SSE	Sustainable Stock Exchange
TCFD	Task Force on Climate-related Financial Disclosure
UN	United Nations
UNCTAD	United Nations Conference on Trade and Development
UNIDO	United Nation Industrial Development Organization
US	United States (of America)
USD	United States Dollar
VRF	Value Reporting Foundation
WBCSD	World Business Council for Sustainable Development

List of Figures

List of Tables

Part I
How to Measure Circularity?—Setting the Foundation

What gets measured gets managed.

—Peter Drucker

The Father of Modern Management

Circularity Assessment: Developing a Comprehensive Yardstick

1

Rashmi Anoop Patil, Sven Kevin van Langen and Seeram Ramakrishna

Abstract

The beginning of this millennium witnessed the emergence of the CE concept with an aim of sustainable development. It gradually gained traction from governments, non-governmental organizations, businesses, and researchers, and the implementation of various strategies towards a CE began. Currently, our economic system is in a transition phase from a linear to a circular one. In this phase, monitoring the progress towards circularity using an assessment framework is of paramount importance, given the critical impact such a transition in the economy may have on the environmental, economic, and social aspects of the society in the coming decades. This work provides an overview of circularity assessment and the four systemic levels of its implementation. The introductory chapter begins with a brief description of how a CE is essential in achieving a sustainable future and provides a glimpse of the current status of our economic system. Further, circularity assessment and the various phases involved in the process are introduced and elaborated. As a highlight of the chapter, we present for the first time, the key features of the upcoming standard from the ISO—the ISO 59020, that aims to establish a generic but optimum process for circularity assessments. Lastly, the chapter concludes with a brief note on the need for understanding the state-of-the-art circularity assessment approaches (as discussed in the following chapters).

R. A. Patil(✉) · S. Ramakrishna(✉)
The Circular Economy Task Force, National University of Singapore, Singapore 117575, Singapore
e-mail: rashmi.anoop33@gmail.com

S. Ramakrishna
e-mail: seeram@nus.edu.sg

S. K. van Langen
UNESCO Chair in Environment, Resources and Sustainable Development (International Ph.D. Programme), Department of Science and Technology, Parthenope University of Naples, 80143 Naples, Italy

Olympia Electronics, Thessaloniki, Greece

S. Ramakrishna
Department of Mechanical Engineering, National University of Singapore, Singapore 117575, Singapore

Centre for Nanotechnology and Sustainability (NUSCNS), 2 Engineering Drive 3, Singapore 117576, Singapore

Keywords

Circularity assessment · Sustainability · Circularity measurement · Circularity indicators · Circularity assessment framework · Circular economy

1.1   Introduction

Our global economy has followed a linear pattern (take-make-use-dispose) [1–3] for around 200 years since the IR. Although the IR has economically benefitted people, businesses, and countries, it has resulted in drawbacks such as accelerated resource extraction, environmental pollution, and excessive consumer demand and consumption. This has led to alarming environmental impacts such as climate change and resource depletion. For instance, globally, material resource use has exceeded 100 billion tons for the first time in history [2, 4]. As a result of global warming, in recent years Greenland's ice sheet is melting much faster than in1990s, leading to rising sea levels [5]. To mitigate such undesired effects, the UN SDGs and the Paris Agreement frameworks came into existence. However, these targets have been yet little achieved, majorly due to slow implementation of policies and inadequate collective action [2, 6–8].

A CE, one that is restorative and regenerative by principle and design, is a practical solution to achieve the desired low-carbon, sustainable economy [3]. In reality, the goal of the Paris Agreement to check global warming to 1.5 °C above the pre-industrial level [9] is achievable by transiting to a CE (due to the reduction in greenhouse gas emissions). Furthermore, CE decouples economic growth from resource constraints and minimizes resource depletion by closing the materials loop. In addition to environmental benefits, chartering towards CE will have indirect advantages such as improved security of raw materials supply, creating new jobs/opportunities in the tertiary sector, and increasing the competitiveness of businesses. According to a McKinsey report [10], the increase in revenue from circular activities (such as the reuse of products and materials, and remanufacturing) together with lower production cost can increase the GDP and thereby facilitate economic growth. In total, the circular approach provides economies an avenue for resilient growth, benefiting all the stakeholders- governments, consumers, businesses, and society as a whole.

1.1.1   Current Landscape

According to the Circularity Gap Report 2020, our current economy is only 8.6% circular and far from reaching the Paris Agreement goal [2, 6]; a clear sign that this is not a sustainable economy. This is primarily due to the classic deep-rooted problems of the wasteful linear economy [3, 11] such as (i) collective contribution of high rates of virgin resource extraction and on-going building-up of stock (in the form of buildings or heavy machinery), (ii) increasing rates of production/manufacturing, and (iii) low rates of processing end-of-life products and cycling back for consumption [2].

In recent years, we have witnessed the CE gaining the attention of researchers, institutions, businesses, and governments in their pursuit of an environmentally, economically, and socially sustainable alternative to the current economic model [3, 12–20]. There have been efforts through multiple fronts such as policies and frameworks [21–23], business models [24–27], research and innovation [28–30], and creating consumer awareness for transiting to a CE [31–34]. Realizing the potential of CE, many countries including those in Europe, and China have introduced CE policies and roadmaps in recent years [19–23]. Also, several global businesses and start-ups have adopted CE principles in their operations and management [35–37]. Even though CE agendas/practices are springing up all around the world, this is merely the beginning, and transiting from a linear to a CE is a gradual and long process.

In this transition phase and beyond, it is crucial to monitor the progress towards circularity and steer the economic system in the right direction. Therefore, the system as a whole and every component of the system needs to be evaluated and feedback should be provided regularly to enhance the circularity. For this purpose, a comprehensive circularity assessment and a standard framework for the same are important.

This chapter introduces the readers to the generic concept of circularity assessment and measurement using a set of indicators. Then, we present the various levels of systemic hierarchy for which circularity assessment is practiced (discussed in detail in the following chapters) prior to discussing the need for a standard circularity assessment framework. Further in the chapter, the details of the International Standards Organization (ISO) 59020—the international standard for circularity assessment, (being developed currently) are furnished. As a concluding opinion, the importance of understanding the state-of-the-art circularity assessment approaches (presented in Chaps. 3–6) and the influence of the upcoming ISO standard on the current practices are presented.

1.2 Circularity Assessment

To assess circularity at a systemic level (such as nation, city, business, and product), it is imperative to measure the relevant environmental, social, and economic indicators. This points to the concept of circularity measurement that can be understood as an approach to determine the circular performance (extent of progress towards a CE) of a systemic level using a set of relevant quantitative and qualitative indicators (see Footnote 1). The collated and processed result from such a measurement process can provide insights on areas of improvement, target timeline, and alternative methods to progress towards circularity. It is to be noted that the data used should be coherent, reliable, traceable, and the procedures/methods for collecting the data and the data sources (inclusive of assumptions) should be available to the users of the assessment framework. Also, a universal metric for measurements should be used so that results can be shared, compared, or reused for other circularity assessments, for example, at a different economic level.

With this background, circularity assessment (see Footnote 1) can be defined as *a process of analyzing and interpreting the the results of circularity measurement, encompassing environmental, economic and social impacts, and balancing significant aspects of the systemic level being assessed such as the target audience, stakeholders' perspectives, application of interpreted data, and complementary assessment methods.*

The CE encompasses the environmental, economic as well as social dimensions and the three dimensions are interdependent at the systemic levels. Therefore, a comprehensive assessment must take into account all the environmental and social impacts (both positive and negative) associated with the subject of assessment (such as product, business, city, region) along with the economic challenges/benefits for achieving circularity. To elaborate, the assessment should not be limited to the circularity indicators for measuring the material resources in the technological cycles as often seen in many proposed methodologies.

1.2.1 Quantifying Circularity Using Indicators

Circularity indicators are crucial instruments for evaluating and communicating the progress of a system towards circularity [38]. In general, indicators can be classified into three categories, namely,

- Quantitative indicators: represented by numerical values that can be used for mathematical calculations and statistical analysis
- Qualitative indicators: descriptive in nature without any quantification
- Semi-quantitative indicators: a qualitative scale that is based on quantitative data

The data set obtained from measuring an indicator is easy to understand as it is a simplified unit of a complex parameter needed to be measured. For example, a complex parameter such as air pollution can be broken down into several indicators for quantifying each of the air pollutants. If an indicator's values are aggregated for a considerably long time (say, a few years), then the accumulated data can be used to observe trends. Therefore, a circularity indicator can be defined as *the basic unit of circularity measurement that translates environmental, social or economic aspects into manageable and understandable data.*

For practical purposes, a comprehensive system of indicators is always of interest for the pursuit of a CE. A circularity indicator system which is a collection of all the circularity indicators should account for all stocks and flows of resources impacting the specified systemic level. The system should also specify how to combine constituting indicators, as well as the methods used to calculate and analyze indicator values in a consistent, replicable manner.

1.2.2 Systemic Levels for Circularity Assessment

In order to ensure proper circularity measurements and assessments appropriate boundaries need to be applied so that the outcome is meaningful and manageable. These boundaries can be either spatial (geographical area) or temporal (time scale) based on the system being considered. In this book, to better understand the circularity implementation and assessment process, the systemic approach for circularity assessment is classified into four levels as listed below (Fig. 1.1).

1. **Macro:** defined by spatial area (e.g., city, country, region, international group) or sector (e.g., mining, manufacturing)
2. **Meso:** defined as group or network of collaborating firms or industries
3. **Micro:** includes individual companies/organizations
4. **Nano:** includes products (inclusive of product as a service) or components reflecting their entire life-cycle

Most prior art on circularity measurement considers only three levels of systemic hierarchy namely, macro, meso, and micro by merging micro and nano levels [13, 14, 18]. However, the 'Circular Metrics, Landscape Analysis' report[1] by the WBCSD mentions four levels of measurement similar to the scale provided here.

1.2.3 Phases of Circularity Assessment

A circularity assessment process involves four phases as shown in Fig. 1.2 (see Footnote 1). These phases are listed below and the generic steps involved in each of them are explained.

1. **Defining Goal and Scope** for the measurement and assessment of circularity:
   - choosing the systemic level and defining its scope
   - selecting the applicable set of circularity indicators
   - establishing data quality requirements
   - pre-selection of complementary methods
   - identifying and establishing communication channels with the stakeholders to document their requirements in the measurement and assessment plan
   - defining specific goals for circularity assessment.

2. **Acquiring Data** for circularity measurement:
   - if possible, deriving data from complementary analysis/standard databases
   - establishing data acquisition protocols with specifications
   - assessing the quality of the acquired data and documenting the same for future reference
   - normalizing the data.

3. **Performing Circularity Measurement:**
   - measuring/calculating indicators values using quantitative data (such as resource flows, impacts)
   - analyzing relevant information to assign values to qualitative indicators
   - performing analyses such as sensitivity, categorical, and relationships as applicable
   - performing simulations/modelling.

4. **Performing Circularity Assessment:**
   - reviewing the objectives and requirements of the circularity assessment
   - analyzing the measurement results to suggest improvements
   - interpreting the measurement results in a specific format for presentation to the target audience
   - answering specific questions to meet the requirements of the stakeholders.

[1] WBCSD's (2018) Circular Metrics Landscape Analysis https://docs. wbcsd.org/2018/06/Circular_Metrics-Landscape_analysis.pdf.

Fig. 1.1 Schematic representation of the four levels of systemic hierarchy considered for circularity assessment—(top-down) macro, meso, micro, and nano. (Right) The scope of each level of the systemic hierarchy is defined. Design adapted from a template; Copyright PresentationGO.com

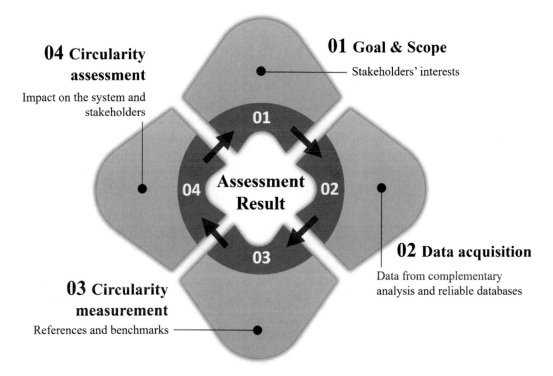

Fig. 1.2 Schematic representation of the four phases of circularity assessment. In the first phase, the goal and scope of the assessment are defined based on the stakeholders' interests, followed by the second phase in which relevant data is acquired from reliable sources. In the third phase, the data is processed using specific techniques, and the total circularity of the system considered is calculated. These results are further analyzed and interpreted to meet the goal of the assessment in the final phase. Based on the initial assessment, the goal and scope can be modified for future iterations. Design adapted from a template; Copyright PresentationGO.com

1.2.4 The Need for a Comprehensive Assessment Framework

There are numerous indicators and proposed methods for measuring circularity at various levels of the economy such as cities or countries, businesses, and products [39–42]. In the past few years, scholars reviewing these have identified the lack of a standard methodology with a comprehensive set of indicators for circularity measurement [42]. This gap has created non-uniform and/or ambiguous measurement frameworks that may lead to incoherent evaluations and/or conclusions. Some of the researchers have highlighted the importance of a well-designed and effective methodology to evaluate the circularity at various levels of the economic hierarchy [41, 42] and have provided recommendations for the development of circularity indicators for efficient evaluation [43–45]. Besides, it is evidenced through extensive studies that none of the existing indices and methodologies are solely capable of monitoring/evaluating the progress towards CE at various systemic levels. Despite the existence of multiple methodologies and indices, the current frameworks for measuring circularity and their application to improve CE strategies are still in their early phase.

At present, it is an established fact that there is a need for a standard methodology such as an ISO framework to assess the circularity of the economic system to aid the transition. Such a framework should be applicable for the circularity assessment of each level of the economic hierarchy—ranging from countries/cities to a product. Monitoring and evaluation framework relying on verifiable data (expressed in standards), with a comprehensive set of performance indicators, and a standard circularity assessment methodology is essential to track the progress towards the CE. This is also significant in avoiding 'lock-in mechanisms' obstructing the progress towards CE [46]. For instance, a situation wherein the non-availability of a certain material (essential in manufacturing a product) in the secondary material market due to economic factors/practical infeasibility of recycling from end-of-life/discarded products leading to the exploitation of rare virgin resource, locks the progress towards CE for a particular industry. A comprehensive circularity assessment framework should provide guidelines and useful feedback for improving the state-of-the-art industrial manufacturing, consumption practices, and existing legislative and policy tools.

1.3 Developing a New Standard for Circularity Assessment: ISO 59020

The ISO has brought together experts around the world to develop the ISO 59000 series of standards for normalizing the understanding of the circular economy and facilitating uniform implementation, monitoring, and measurement of circularity. This series is categorized into several standards as listed below.

- ISO 59004: Circular Economy—Terminology, Principles, and Guidance for Implementation
- ISO 59010: Circular Economy—Guidance on Business Models and Value Networks
- ISO 59014: Secondary materials—Principles, Sustainability, and traceability requirements
- ISO 59020: Circular Economy—Measuring and Assessing Circularity
- ISO 59031: Circular Economy—Performance-based Approaches
- ISO 59040: Circular Economy—Product Circularity Data Sheet.

In this chapter, we present the ISO 59020 on circularity assessment, which includes the assessment at all the systemic levels in response to the market needs. As of February 2022, the draft of the ISO 59020 standard is being developed, discussed, and modified to cater to the interests of all the stakeholders.

1.3.1 Principles for ISO 59020

Before discussing the details of the upcoming ISO model, one needs to understand the principles based on which the ISO framework is being developed. The twelve principles considered for this standard are as listed below and are documented in the ISO 59020 draft version 1.0.[2]

1. **Applicability:** The circularity measurement (using a system of indicators) and assessment should be applicable to every systemic level and across various sectors.

2. **Coherence:** The circularity assessment method should be based on the defined metrics to achieve coherent and reproducible results.

3. **Comparability:** The measurement metrics and the assessment results should enable reliable comparison of two or more similar entities belonging to the same systemic level.

4. **Completeness:** The assessment should account for all stocks and flows, the environmental, social and economic impacts of the system under consideration. For example, in the case of a product, from planning and design, through the selection of raw materials, manufacturing, operations and processes, distribution, use, maintenance, reuse, repair, recycling, or other circular practices, inclusive of accounting for mass and values along the chain, to any losses such as final dispositions and emissions (as documented in the ISO 59020 draft version 1.0 (see Footnote 1)).

5. **Reliability of data:** The data used or provided for the assessment should reflect the best available know-how from measurement or calculation and should come with a descriptor such as 'quantified measurement', 'verified' to 'non-qualified estimate' (as documented in the ISO 59020 draft version 1.0). The data used should be referenced and traceable, and be as complete and consistent as reasonably possible (e.g. from well-developed data sets and well-maintained databases). Any data sources of lower quality used or assumptions made in an analysis should be carefully managed and reported along with (i) a quantitative assessment of the resulting uncertainties, (ii) the identification of data sources or assumptions that represent key sensitivities in

the analysis and (iii) the potential uncertainty in the overall result (see Footnote 1).

6. **Loss rate estimations:** Circularity assessments should include calculation of losses from the system being assessed. Losses may include emissions, expended energy, disappearances and other changes in resource values associated with each step or activity within the system, for example, breakages, lost materials and products, costs of collection, transportation, process inefficiencies, and even consumer behavior.

7. **Robustness:** The impact of uncertainties in data used for assessment should be evaluated with standard tools (for example, sensitivity analysis and comparison of analysis using different indicator systems and databases). The interpretation of the circularity assessment and the related conclusions should not fundamentally change as a result of uncertainties in the data used.

8. **Scale universality:** The assessment method and the indicator system used should be applicable to any scale in the systemic hierarchy (see Footnote 1) such as:
   (i) from single unit sales and trade transactions to global sales and supply chains up to whole economies,
   (ii) from short term to multigeneration changes that result from a change in approach, and
   (iii) from small user applications to governmental and global policy decisions.

9. **Systemic interdependencies:** An assessment should take into account all the relevant systematic interdependencies, both internal and external to the scope of assessment. Internal interdependencies might be represented by a matrix of relationships between data and datasets, methodologies, and indicators. Interdependencies external to the scope of the level of assessment should also be considered (for example, long-term impacts, changes in customer behavior, effects on a level outside the assessment level, impacts of one country's circularity measures on another). Also, interdependencies between the three aspects of sustainability (environmental, economic,

---

[2] Preliminary draft of ISO 59020, Version: 1.0 by ISO/TC 323 Working Group 3 for 'Circular Economy—Measuring Circularity'.

and social) should be recognized (see Footnote 1).

10. **Traceability of resources:** The traceability of resources and materials within the scope of circularity assessment is critical. It is the ability to follow the history and future of the resources and materials used as well as the products resulting from these so that the circularity of a system can be practically verified. Materials flow analysis and life cycle analysis are important tools in identifying the usage and fate of the resources. Traceability must extend to cover all aspects that impinge on the extraction of resources from the natural environment or their reintroduction to the system as well as any positive or negative impacts such activities may have on the short or long term regeneration of biological systems (including soil and water) and indigenous ecosystems (see Footnote 1).

11. **Transparency:** The methods, models, procedures, and data sources used in a circularity assessment should be transparent and unambiguous. The assessment process should be available to all interested parties to the maximum possible extent; taking account of confidentiality where appropriate. Consistent documentation of data; collection, calculation, and reporting of data by specifying and structuring relevant information ensures transparency and permits comparative analysis. Uncertainty or volatility in data, estimations, and assumptions made during the assessment should be declared. Where uncertainties exist or assumptions have been made, sufficient data should be provided in any analysis or results to enable the third-party calculation of alternative scenarios following the documented assumptions or applying different datasets (see Footnote 1).

12. **Time scale:** The time scale/temporal dimension of data in an assessment should be properly considered and appropriately detailed. Specifically, factors such as product lifetime, the frequency with which materials, parts, components, and products enter the loop through reuse, sharing, repair, refurbishment, remanufacturing, and recycling,

and the expected time to end-of-life should be provided. The time scale chosen should encompass the entire life cycle and resource recovery.

1.3.2 The ISO 59020 Model

It is to be noted that at the time of authoring this book (Feb 2022), the ISO 59020 is under development and the schematic model (Fig. 1.3) of the framework provided here may differ from the final version.[3] The intention of developing this framework is to provide a standard and comprehensive circularity assessment methodology that is applicable at every systemic level catering to the interests of the related stakeholders. Based on the other established standards that are complementary to the this (such as 14000 series), we can expect the standard to include a thorough and comprehensive inventory to measure the energy associated with the systemic level considered, along with the material circularity. Thus, it can be a reliable methodology to provide the cummulative value (for the extent of circularity) and insights to improve the overall circularity profile of the systemic level in consideration. However, the standard may need regular updations and amends based on users' feedback.

As discussed earlier in Sect. 1.2.3, the model consists of 4 major stages, namely, (i) Goal and Scope, (ii) Circularity Measurement, (iii) Data Acquisition, and (iv) Circularity Assessment. In the Fig. 1.3, these stages are shown in the gray boxes within the framework. The dotted box is for complementary assessment methods/standard databases which can aid the measurement process. The arrows indicate the information flow between the stages they connect.

In the first stage of circularity assessment framework (see Footnote 2), the boundaries of the systemic level considered, and the requirements of the stakeholders are defined. Based on these inputs, the relevant indicators (including environmental, social and economic factors) are selected from the inventory and the circularity is measured using pre-determined protocols/guidelines, accordingly. Besides, the data required for the

---

[3]ISO 59020 CD Living document #2, version 23 February 2022 (Draft of framework for measuring and assessing circularity) on 'Circular Economy—Measuring and Assessing Circularity'.

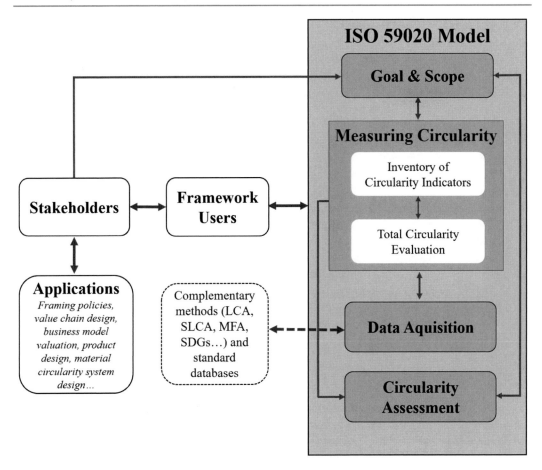

Fig. 1.3 Schematic conceptual model of the ISO 59020. The model consists of 4 major stages, namely, (i) Goal and Scope, (ii) Circularity Measurement, (iii) Data Acquisition, and (iv) Circularity Assessment as shown in the gray boxes within the framework. The measurement process has provisions to borrow data from complementary methods. The framework can be used by analysts to derive insights into the circular performance of the system considered, and assist the stakeholders in their efforts to enhance the circularity of the system. Reproduced with permission from ISO 59020 CD Living document #2, version 23 February 2022 (Draft of framework for measuring and assessing circularity) on 'Circular Economy—Measuring and Assessing Circularity' (see Footnote 2)

indicators measurements can be borrowed from complementary methods/standard databases. This provision to use complementary methods such as Life-cycle Assessment (LCA), Materials Flow Analysis (MFA), SDGs evaluation method, and related ISO frameworks (for example, ISO 14000 series, ISO 26000, etc.) for data collection is allowed since the measurement framework is new. Specific algorithms in the data acquisition stage are used to normalize and process all the data, to calculate the total circularity of the system. Finally, the results of measurement are analyzed and interpreted to present them in an understandable form. Analysts and researchers (framework users) who evaluate/assess the overall circularity based on the assessment framework can provide their insights on the progress towards a CE at the systemic level and indicate the specific areas for improvement as per the requirements of the stakeholders. These insights are useful in setting the rules and regulations, enframing policies and legislation, business remodeling, circular product design, and enhancing the circularity of the value chain.

1.3.3 Limitations of Circularity Assessment

Although care is taken to develop a robust assessment framework, there will always be several challenges in its implementation. Such extensive assessments involve several variables and uncertainties, limiting the efficiency of the assessment process. Some of the major limitations of circularity assessment are listed below (see Footnote 1).

1. A circularity assessment result obtained from applying the framework to a systemic component cannot be used to make claims on the circular performance of another component in the same systemic level even though they are similar. For instance, the assessment result of an automobile manufacturing company cannot be used to make assumptions about the performance of another company belonging to the same industrial sector.

2. Though the assessment results from a lower level of the systemic hierarchy are transferable to a higher level (for example, from nano to macro), they may not be taken into account properly (as in partial usage of assessment data) resulting in improper assessment of the higher systemic level.

3. The circularity performance of two systemic components belonging to the same level cannot be compared if they are assessed using different indicator systems or subsets of the same indicator system (based on the goal and scope of each) even if the same resources are studied.

4. Lack in the availability of appropriate data for each indicator considered and the lack of spatial and temporal dimensions during data collection introduces uncertainty to the measurement and assessment process.

5. The unknown and unpredictable environmental and social impacts and the economic impacts reflecting the market changes may dynamically affect the assessment results.

1.4 Relevance of This Work

While the ISO framework for circularity assessment is under development, it is important to understand the state-of-the-art circularity assessment at various levels of systemic hierarchy. This will provide an overall perspective of the advantages and issues related to the current practices. These crucial insights may not only aid in minimizing the uncertainties/limitations in the current practices but also assist in improving the framework and its implementation.

In this regard, a detailed discussion on the circularity assessment carried out currently at each level of the systemic hierarchy is provided from Chaps. 3 to 6. Each chapter provides an overview of the assessment process for a particular systemic level with relevant set of indicators, along with real-world cases for a better understanding of the implementation of such a process. It is the authors' opinion that consumers being a crucial part of the economic system can influence the transition towards circularity profoundly. Hence, as a befitting conclusion to this book, the role of consumers (considered as the granular systemic level) in the pursuit of a circular society is discussed, and how their behavior can affect a ripple effect of change in the economic system as a whole.

References

1. Sariatli F (2017) Linear economy versus circular economy: a comparative and analyzer study for optimization of economy for sustainability. Visegrad Journal on Bioeconomy and Sustainable Development 6(1):31–34
2. Wit, Mde, Hoogzaad, J, Daniels, Cvon (2020) The circularity gap report 2020
3. Ellen MacArthur Foundation (2013) Towards the circular economy: economic and business rationale for an accelerated transition
4. IRP (2017) Assessing global resource use: a systems approach to resource efficiency and pollution reduction
5. Shepherd A, Ivins E, Rignot E et al (2020) Mass balance of the Greenland ice sheet from 1992 to 2018. Nature 579(7798):233–239

6. Rogelj J, Den Elzen M, Höhne N et al (2016) Paris agreement climate proposals need a boost to keep warming well below 2 °C. Nature 534(7609):631–639

7. Allen C, Metternicht G, Wiedmann T (2018) Initial progress in implementing the sustainable development goals (SDGs): a review of evidence from countries. Sustain Sci 13(5):1453–1467

8. von Stechow C, Minx JC, Riahi K et al (2016) 2 °C and SDGs: united they stand, divided they fall? Environ Res Lett 11(3):034,022

9. Schleussner CF, Rogelj J, Schaeffer M et al (2016) Science and policy characteristics of the Paris agreement temperature goal. Nat Clim Chang 6(9):827–835

10. McKinsey Center for Business and Environment (2015) Growth within: a circular economy vision for a competitive Europe

11. Ramkumar S, Kraanen F, Plomp R et al (2018) Linear risks (Joint project between Circle Economy, PGGM, KPMG, EBRD, and WBCSD)

12. Stahel WR (2016) The circular economy. Nat News 531(7595):435

13. Ghisellini P, Cialani C, Ulgiati S (2016) A review on circular economy: the expected transition to a balanced interplay of environmental and economic systems. J Clean Prod 114:11–32

14. Kirchherr J, Reike D, Hekkert M (2017) Conceptualizing the circular economy: an analysis of 114 definitions. Resour Conserv Recycl 127:221–232

15. Murray A, Skene K, Haynes K (2017) The circular economy: an interdisciplinary exploration of the concept and application in a global context. J Bus Ethics 140(3):369–380

16. Geissdoerfer M, Savaget P, Bocken NM et al (2017) The circular economy-a new sustainability paradigm? J Clean Prod 143:757–768

17. Korhonen J, Honkasalo A, Seppälä J (2018) Circular economy: the concept and its limitations. Ecolog Econ 143:37–46

18. Corona B, Shen L, Reike D et al (2019) Towards sustainable development through the circular economy - a review and critical assessment on current circularity metrics. Resour Conserv Recycl 151(104):498

19. Salvatori G, Holstein F, Böhme K (2019) Circular economy strategies and roadmaps in Europe: identifying synergies and the potential for cooperation and alliance building

20. Yong R (2007) The circular economy in china. J Mater Cycles Waste Manag 9(2):121–129

21. McDowall W, Geng Y, Huang B et al (2017) Circular economy policies in China and Europe. J Ind Ecol 21(3):651–661

22. Zhu J, Fan C, Shi H et al (2019) Efforts for a circular economy in China: a comprehensive review of policies. J Ind Ecol 23(1):110–118

23. Hartley K, van Santen R, Kirchherr J (2020) Policies for transitioning towards a circular economy: expectations from the European Union (EU). Resour Conserv Recycl 155(104):634

24. Tse T, Esposito M, Soufani K (2016) How businesses can support a circular economy. Harv Bus Rev 30

25. Lewandowski M (2016) Designing the business models for circular economy-towards the conceptual framework. Sustainability 8(1):43

26. Bocken NM, De Pauw I, Bakker C et al (2016) Product design and business model strategies for a circular economy. J Ind Prod Eng 33(5):308–320

27. Lüdeke-Freund F, Gold S, Bocken NM (2019) A review and typology of circular economy business model patterns. J Ind Ecol 23(1):36–61

28. Keijer T, Bakker V, Slootweg JC (2019) Circular chemistry to enable a circular economy. Nat Chem 11(3):190–195

29. Virtanen M, Manskinen K, Eerola S (2017) Circular material library. an innovative tool to design circular economy. Des J 20(sup1):S1611–S1619

30. Mestre A, Cooper T (2017) Circular product design. a multiple loops life cycle design approach for the circular economy. Des J 20(sup1):S1620–S1635

31. Antikainen M, Lammi M, Paloheimo H et al (2015) Towards circular economy business models: consumer acceptance of novel services. In: Proceedings of the ISPIM innovation summit, Brisbane, Australia, pp 6–9

32. Hazen BT, Mollenkopf DA, Wang Y (2017) Remanufacturing for the circular economy: an examination of consumer switching behavior. Bus Strat Environ 26(4):451–464

33. Kuah AT, Wang P (2020) Circular economy and consumer acceptance: an exploratory study in East and Southeast Asia. J Clean Prod 247(119):097

34. Sijtsema SJ, Snoek HM, Van Haaster-de Winter MA et al (2020) Let's talk about circular economy: a qualitative exploration of consumer perceptions. Sustainability 12(1):286

35. Aranda-Usón A, Portillo-Tarragona P, Scarpellini S et al (2020) The progressive adoption of a circular economy by businesses for cleaner production: an approach from a regional study in Spain. J Clean Prod 247(119):648

36. Shirvanimoghaddam K, Motamed B, Ramakrishna S et al (2020) Death by waste: fashion and textile circular economy case. Sci Total Environ 718(137):317

37. Patil RA, Ghisellini P, Ramakrishna S (2021) Towards sustainable business strategies for a circular economy: environmental, social and governance (ESG) performance and evaluation. In: An introduction to circular economy, Springer, pp 527–554

38. Gabrielsen P, Bosch P (2003) Environmental indicators: typology and use in reporting. EEA, Copenhagen

39. Mayer A, Haas W, Wiedenhofer D et al (2019) Measuring progress towards a circular economy: a monitoring framework for economy-wide material loop closing in the EU28. J Ind Ecol 23(1):62–76

40. Moraga G, Huysveld S, Mathieux F et al (2019) Circular economy indicators: what do they measure? Resour Conserv Recycl 146:452–461

41. Saidani M, Yannou B, Leroy Y et al (2019) A taxonomy of circular economy indicators. J Clean Prod 207:542–559

42. Kristensen HS, Mosgaard MA (2020) A review of micro level indicators for a circular economy-moving away from the three dimensions of sustainability? J Clean Prod 243(118):531

43. Saidani M, Yannou B, Leroy Y et al (2017) How to assess product performance in the circular economy? Proposed requirements for the design of a circularity measurement framework. Recycling 2(1):6

44. Niero M, Kalbar PP (2019) Coupling material circularity indicators and life cycle based indicators: a proposal to advance the assessment of circular economy strategies at the product level. Resour Conserv Recycl 140:305–312

45. Iacovidou E, Velis CA, Purnell P et al (2017) Metrics for optimising the multi-dimensional value of resources recovered from waste in a circular economy: a critical review. J Clea Prod 166:910–938

46. Salmenperä H (2021) Different pathways to a recycling society-comparison of the transitions in Austria, Sweden and Finland. J Clean Prod 292(125):986

Insights from ESG Evaluation for Circularity Assessment

2

Sven Kevin van Langen, Rashmi Anoop Patil
and Seeram Ramakrishna

Abstract

Sustainability and circularity co-exist and both encompass environmental, social, and economic aspects. In the past few years, we have witnessed ESG reporting gaining traction and with it, a fast rise in the market for sustainable finance and development. Developed countries are coming up with regulations governing the ESG reporting standards and performance, which has been a voluntary action so far. Several ESG reporting standards have been developed across the globe, typically containing many metrics related to circularity in both environmental and social aspects. In this chapter, several such global standards are discussed and one national implementation by the Athens stock exchange is detailed. Later in this chapter, current and upcoming regulations regarding ESG reporting in developed countries are provided. The European Union is the most advanced body of regulation on this topic and is covered in more detail. Trends in the sustainable finance and bonds markets are also presented at the end of this chapter, a market that will finance projects and developments towards circularity and sustainability. The chapter concludes with a call to incorporate a clear circularity assessment within ESG reports. Standard definitions of practices considered circular and how to best measure them need to be further developed.

Keywords

ESG · Circularity assessment · Sustainable finance · Circular businesses · Sustainability reporting

S. K. van Langen(✉)
UNESCO Chair in Environment, Resources and Sustainable Development (International Ph.D. Programme), Department of Science and Technology, Parthenope University of Naples, 80143 Naples, Italy
e-mail: kevin.vanlangen@studenti.uniparthenope.it

Olympia Electronics, Thessaloniki, Greece

R. A. Patil · S. Ramakrishna(✉)
The Circular Economy Task Force, National University of Singapore, Singapore 117575, Singapore
e-mail: seeram@nus.edu.sg

S. Ramakrishna
Department of Mechanical Engineering, National University of Singapore, Singapore 117575, Singapore

Centre for Nanotechnology and Sustainability (NUSCNS), 2 Engineering Drive 3, Singapore 117576, Singapore

© The Author(s) 2023
R. A. Patil and S. Ramakrishna (eds.), *Circularity Assessment: Macro to Nano*,
https://doi.org/10.1007/978-981-19-9700-6_2

2.1   Introduction

Where Chap. **??** introduced the generic concept of circularity assessment and its standardization, this chapter will consider circularity assessment within the wider framework of ESG reporting. ESG reporting finds its origin in a stronger push for CSR, specifically the will to embed non-financial responsibilities in a company's governance and to create more ethical business models [1]. Especially the environmental category of ESG reporting metrics typically contains information related to a firm's sustainability. An UNCTAD report showed that over half of the global top 100 listed companies report on CO_2 emissions, water consumption, waste generation, the reuse of waste, and energy consumption in their ESG reporting [2].

CSR strategies were originally considered as a tool to repair a tarnished brand name. The concept evolved a lot in the earliest 21st century when the financial sector faced a lot of scandals [1]. Over time, it proved that a good CSR performance led to increased access to financial markets and enhanced financial performance of firms, and consequently saw a broader adoption as well as gained governments' interest [3, 4]. Recent studies show similar results for ESG reporting [5].

If ESG indicators have a higher strategic relevance, investors are more willing to invest in a company [6]. According to a study on companies that are enlisted in the Hong Kong stock exchange, ESG initiatives are better received by the market than other sustainability initiatives [7]. Especially, socially responsible investors are greatly enabled by good ESG reporting [8], often serving as a proxy for sustainability scores. One particular study in Japan found that ESG policy adoption creates indirect value creation for companies by first increasing their capacity to innovate, which in turn increases that company's financial performance [9]. Listed companies with a high market capitalization benefit more financially from ESG disclosure than companies with a smaller market capitalization [10, 11]. In contrast, Chinese A-share firms do not show significant positive financial performance with a good ESG rating [12]. In another developing country, Bangladesh, there

was a positive correlation [13]. ESG disclosure is most crucial for short-term profits while taking action to improve ESG ratings is more important for long-term financial performance [14].

In addition to ESG metrics correlating with a firm's financial and economic performance (with a better ESG performance generally leading to a better financial and economic performance), the metrics also have strong interdependencies amongst the ESG metrics themselves, with environmental, social, and governance metrics enforcing each other [15, 16]. These interdependencies differ sector-wise. For example, for the consumer non-cyclical, healthcare, technology, telecommunications, and utility sectors, it is better to mainly focus on social performance. For basic materials, consumer cyclicals, and financial sectors, it is best to focus on environmental performance first to get greater overall sustainability [15]. The travel and leisure industry and the pharmaceutical industry seem to benefit most from a good governance performance [17, 18].

Despite the increasing evidence that ESG reporting has positive economic and financial effects on companies, it is still not exactly clear how to best motivate firms to report their ESG performance. S&P 500 companies, i.e., 500 of the largest companies listed on stock exchanges in the US, tend to differ a lot in their levels of disclosure [11]. S&P 500 firms are most likely to disclose corporate governance information and least likely to disclose environmental information (including CE metrics). One study in Italy found that high diversity in the board of directors has a significant correlation with ESG disclosure [19]. Having women on the board of directors shows a positive correlation in a majority of studies with large sample sizes [11, 19–21]. Linking executive compensation to ESG performance also shows a correlation with better ESG disclosure [11] (Fig. 2.1).

In the following section, efforts taken towards standardizing ESG reports amongst firms and the potential of aggregating ESG reports useful for macro-analysis and policymaking are discussed. To further push ESG reporting, and indirectly, sustainable finance, governments have started to enforce ESG disclosure through regulations,

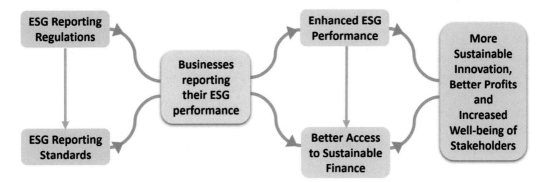

Fig. 2.1 An overview of the relationship between ESG regulations and standards (for reporting), ESG performance of businesses, and sustainable finance. The way businesses report their ESG performance is dependent on the government regulations and the standards set by stock exchanges and/or other organizations. The reporting also provides insights to the business stakeholders to improve their ESG performance and consequently get better access to sustainable finance. This can lead to more innovation and better profits. Design adapted from a template; Copyright PresentationGO.com

which are discussed next. ESG metrics suffer from a lack of transparency and do not show enough convergence [8]. The lack of transparency leads to 'green washing', meaning that companies define products and processes as green/sustainable while in reality, they aren't. The lack of convergence makes it harder to compare the available reports. The growth of the sustainable finance market, a market heavily reliant on ESG reporting which is growing at a rapid rate, is discussed. The chapter concludes with a summary of the lessons learned in this chapter.

2.2 Non-financial Reporting Standards

As previously mentioned, there are no universal standards for ESG reporting. Currently, regulated standards are being developed in the EU [22, 23]. Firms do have some established guidelines to adhere to, specifying what metrics to report on. In this section, some of the established international guidelines and one nation-specific reporting guide provided by the Athens stock exchange are discussed.

2.2.1 International ESG Reporting Guidelines

2.2.1.1 The GRI

The GRI is one of the most used reporting frameworks. Through better support to organizations, the initiative aims to make sustainability reporting a standard practice. The GRI provides the guidelines as two sets—(i) universal standards (set of basic metrics) and (ii) topic-specific standards (economic, social, and environmental metrics) to choose as per their needs of reporting.[1] Furthermore, GRI provides sector-specific standards. The disclosure topics covered by GRI are provided in Table 2.1. The GRI standards are continuously updated. For this reason, the standards have a modular approach to make it easier for companies as they do not have to completely overhaul their reporting when new standards are published. However, this does lead to increased discrepancy between the reports of different companies. The GRI tries to make the guidelines universally applicable to large and small, public and private companies. It provides such basic metrics with the aim that every company can report on them and more in-depth metrics for companies that have more resources to spend on sustainability reporting. The metrics are meant to better support internal organization policies and strategies as well as to help external stakeholders to evaluate an organization.

---

[1] GRI https://www.globalreporting.org.

Table 2.1 List of disclosures provided by the GRI. Adapted from globalreporting.org

GRI STANDARD	DISCLOSURE
GRI 2: General Disclosures 2021	2-1 Organizational details
	2-2 Entities included in the organization's sustainability reporting
	2-3 Reporting period, frequency and contact point
	2-4 Restatements of information
	2-5 External assurance
	2-6 Activities, value chain and other business relationships
	2-7 Employees
	2-8 Workers who are not employees
	2-9 Governance structure and composition
	2-10 Nomination and selection of the highest governance body
	2-11 Chair of the highest governance body
	2-12 Role of the highest governance body in overseeing the management of impacts
	2-13 Delegation of responsibility for managing impacts
	2-14 Role of the highest governance body in sustainability reporting
	2-15 Conflicts of interest
	2-16 Communication of critical concerns
	2-17 Collective knowledge of the highest governance body
	2-18 Evaluation of the performance of the highest governance body
	2-19 Remuneration policies
	2-20 Process to determine remuneration
	2-21 Annual total compensation ratio
	2-22 Statement on sustainable development strategy
	2-23 Policy commitments
	2-24 Embedding policy commitments
	2-25 Processes to remediate negative impacts
	2-26 Mechanisms for seeking advice and raising concerns
	2-27 Compliance with laws and regulations
	2-28 Membership associations
	2-29 Approach to stakeholder engagement
	2-30 Collective bargaining agreements
GRI 3: Material Topics 2021	3-1 Process to determine material topics
	3-2 List of material topics
	3-3 Management of material topics
GRI 201: Economic Performance 2016	201-1 Direct economic value generated and distributed
	201-2 Financial implications and other risks and opportunities due to climate change
	201-3 Defined benefit plan obligations and other retirement plans
	201-4 Financial assistance received from government

(continued)

Table 2.1 (continued)

GRI STANDARD	DISCLOSURE
GRI 202: Market Presence 2016	202-1 Ratios of standard entry level wage by gender compared to local minimum wage
	202-2 Proportion of senior management hired from the local community
GRI 203: Indirect Economic Impacts 2016	203-1 Infrastructure investments and services supported
	203-2 Significant indirect economic impacts
GRI 204: Procurement Practices 2016	204-1 Proportion of spending on local suppliers
GRI 205: Anti-corruption 2016	205-1 Operations assessed for risks related to corruption
	205-2 Communication and training about anti-corruption policies and procedures
	205-3 Confirmed incidents of corruption and actions taken
GRI 206: Anti-competitive Behavior 2016	206-1 Legal actions for anti-competitive behavior, anti-trust, and monopoly practices
GRI 207: Tax 2019	207-1 Approach to tax
	207-2 Tax governance, control, and risk management
	207-3 Stakeholder engagement and management of concerns related to tax
	207-4 Country-by-country reporting
GRI 301: Materials 2016	301-1 Materials used by weight or volume
	301-2 Recycled input materials used
	301-3 Reclaimed products and their packaging materials
GRI 302: Energy 2016	302-1 Energy consumption within the organization
	302-2 Energy consumption outside of the organization
	302-3 Energy intensity
	302-4 Reduction of energy consumption
	302-5 Reductions in energy requirements of products and services
GRI 303: Water and Effluents 2018	303-1 Interactions with water as a shared resource
	303-2 Management of water discharge-related impacts
	303-3 Water withdrawal
	303-4 Water discharge
	303-5 Water consumption
GRI 304: Biodiversity 2016	304-1 Operational sites owned, leased, managed in, or adjacent to, protected areas and areas of high biodiversity value outside protected areas
	304-2 Significant impacts of activities, products and services on biodiversity
	304-3 Habitats protected or restored
	304-4 IUCN Red List species and national conservation list species with habitats in areas affected by operations

(continued)

Table 2.1 (continued)

GRI STANDARD	DISCLOSURE
GRI 305: Emissions 2016	305-1 Direct (Scope 1) GHG emissions
	305-2 Energy indirect (Scope 2) GHG emissions
	305-3 Other indirect (Scope 3) GHG emissions
	305-4 GHG emissions intensity
	305-5 Reduction of GHG emissions
	305-6 Emissions of ozone-depleting substances (ODS)
	305-7 Nitrogen oxides (NOx), sulfur oxides (SOx), and other significant air emissions
GRI 306: Waste 2020	306-1 Waste generation and significant waste-related impacts
	306-2 Management of significant waste-related impacts
	306-3 Waste generated
	306-4 Waste diverted from disposal
	306-5 Waste directed to disposal
GRI 308: Supplier Environmental Assessment 2016	308-1 New suppliers that were screened using environmental criteria
	308-2 Negative environmental impacts in the supply chain and actions taken
GRI 401: Employment 2016	401-1 New employee hires and employee turnover
	401-2 Benefits provided to full-time employees that are not provided to temporary or part-time employees
	401-3 Parental leave
GRI 402: Labor/Management Relations 2016	402-1 Minimum notice periods regarding operational changes
GRI 403: Occupational Health and Safety 2018	403-1 Occupational health and safety management system
	403-2 Hazard identification, risk assessment, and incident investigation
	403-3 Occupational health services
	403-4 Worker participation, consultation, and communication on occupational health and safety
	403-5 Worker training on occupational health and safety
	403-6 Promotion of worker health
	403-7 Prevention and mitigation of occupational health and safety impacts directly linked by business relationships
	403-8 Workers covered by an occupational health and safety management system
	403-9 Work-related injuries
	403-10 Work-related ill health
GRI 404: Training and Education 2016	404-1 Average hours of training per year per employee
	404-2 Programs for upgrading employee skills and transition assistance programs
	404-3 Percentage of employees receiving regular performance and career development reviews

(continued)

Table 2.1 (continued)

GRI STANDARD	DISCLOSURE
GRI 405: Diversity and Equal Opportunity 2016	405-1 Diversity of governance bodies and employees
	405-2 Ratio of basic salary and remuneration of women to men
GRI 406: Non-discrimination 2016	406-1 Incidents of discrimination and corrective actions taken
GRI 407: Freedom of Association and Collective Bargaining 2016	407-1 Operations and suppliers in which the right to freedom of association and collective bargaining may be at risk
GRI 408: Child Labor 2016	408-1 Operations and suppliers at significant risk for incidents of child labor
GRI 409: Forced or Compulsory Labor 2016	409-1 Operations and suppliers at significant risk for incidents of forced or compulsory labor
GRI 410: Security Practices 2016	410-1 Security personnel trained in human rights policies or procedures
GRI 411: Rights of Indigenous Peoples 2016	411-1 Incidents of violations involving rights of indigenous peoples
GRI 413: Local Communities 2016	413-1 Operations with local community engagement, impact assessments, and development programs
	413-2 Operations with significant actual and potential negative impacts on local communities
GRI 414: Supplier Social Assessment 2016	414-1 New suppliers that were screened using social criteria
	414-2 Negative social impacts in the supply chain and actions taken
GRI 415: Public Policy 2016	415-1 Political contributions
GRI 416: Customer Health and Safety 2016	416-1 Assessment of the health and safety impacts of product and service categories
	416-2 Incidents of non-compliance concerning the health and safety impacts of products and services
GRI 417: Marketing and Labeling 2016	417-1 Requirements for product and service information and labeling
	417-2 Incidents of non-compliance concerning product and service information and labeling
	417-3 Incidents of non-compliance concerning marketing communications
GRI 418: Customer Privacy 2016	418-1 Substantiated complaints concerning breaches of customer privacy and losses of customer data

2.2.1.2 The Value Reporting Foundation (VRF)

In June 2021, the IIRC merged with the SASB into the VRF. As the results of this merger are yet to materialize, we will discuss the IIRC and the SASB separately. The IIRC provides guiding principles for reporting, while the SASB provides industry-specific disclosure topics. Taken together, the VRF hopes to harmonize the reporting processes and create comparability between reports.

The IIRC provides a framework that helps develop integrated reporting, linking financial, manufacturing, human, intellectual, natural, relational, and social capital. It aims to furnish better information to providers of financial capital, promote more coherent and efficient reporting, enhance accountability and stewardship, understand the interdependencies between different types of capital, and support integrated thinking from the short to the long term. The latest version of this framework (at the time of writing this chapter) has been released in January 2021, this was the first revision since the original release in 2013.

The SASB is a US-based NGO that provides industry-specific metrics for sustainability reporting. Currently, SASB covers 77 industries and provides the ESG metrics most relevant for financial performance in each of these industries. The focus of SASB on metrics that affect financial performance most, is what sets SASB apart from GRI. SASB metrics are important for investors and integrate well with financial reporting.

2.2.1.3 Carbon Disclosure Project (CDP)

The CDP provides a set of questions for companies to assess their progress in minimizing climate change risks under the four categories: climate change, supply chain, water usage, and forestry management services [1]. The CDP aims specifically at companies, cities, and states or regions, and tries to also help its clients to align with regulations and the G20's Task Force on climate-related financial disclosures. Standards for measuring carbon footprint relevant for ESG reporting are provided by the CDP.

2.2.1.4 Task Force on Climate-Related Financial Disclosure (TCFD)

TCFD is a voluntary and market-driven initiative [2]. It provides recommendations to improve the disclosure of climate-related information in existing financial reports. TCFD's objective is to help companies better identify risks and opportunities that come from climate change, which ultimately helps investors to make better decisions. The recommendations are categorized into the following four areas: governance, strategy, risk management, metrics, and targets. TCFD also provides further principles and recommends climate-change scenario analysis.

2.2.2 The ATHEX ESG Reporting Guide

In recent times, many stock exchanges have chosen to self-regulate the ESG disclosures of the listed companies and provide a voluntary reporting guide on what metrics to report. The ATHEX [3] is one of them, inspired by the UN SSE initiative in 2018. The guide offers both a suggestion of metrics to report on and offers practical guidelines on how to disclose them. Other stock exchanges around the world offer similar guidelines. The objectives of ATHEX ESG Reporting guidelines[2] are as listed below.

- To create more awareness on why ESG disclosure is important and highlight opportunities arising from it.
- To make it easier for companies to disclose ESG information.
- To improve the quality, comparability, and availability of ESG information provided by firms listed with ATHEX.
- To enhance the image of Greek companies.
- To improve information flows between companies, investors, and other stakeholders.
- To help investors better utilize ESG data in their decision-making.

---

[2] ATHEX guidelines https://www.athexgroup.gr/esg-reporting-guide.

While the reporting guide is mainly aimed at companies listed with ATHEX, it also aims to be relevant for other companies, small and big, and in all sectors that want to voluntarily disclose ESG data. It provides best practices for both companies that already do ESG reporting and companies that want to do ESG reporting for the first time. While the guide is for voluntary reporting, it aligns with requirements from the EU's active Non-Financial Reporting Directive (which applies to Greece as the country is an EU member). The metrics provided are increasingly important to investors for both listed and non-listed companies. The expected results of following the guidelines, according to ATHEX, are as follows:

- Better access to capital
- Compliance with future regulatory changes
- Better corporate performance
- Better stakeholder engagement
- A better reputation.

2.2.2.1 Core Metrics

The ATHEX guide provides some core metrics (listed in Table 2.2) that it advises for all companies. They are deemed the most important metrics and universal across the entire economy and applicable to companies in all sectors.

2.2.2.2 Advanced Metrics

The advanced metrics (listed in Table 2.3) should help companies that perform well in ESG areas to showcase their good performance, attracting more investments.

2.2.2.3 Sector-Specific Metrics

Sector-specific metrics are not relevant to all companies, but only to a subset of companies listed with ATHEX. These metrics are listed in Table 2.4 mostly highlight ESG-related risks that are relevant to specific sectors (e.g., the management approach to the use of critical materials for the technology and communication sector or environmental and social management in manufacturing companies).

2.3 Non-financial Reporting Regulations in Different Regions of the World

This section will detail several regulations related to ESG reporting that currently exist or are in development. Harmonized regulation on both ESG reporting, ESG performance, and defining circular practices will lead to a further thrust for the adoption of the CE, as investors and firms get more security on their investments. Special focus is given to regulation in Europe as Europe is by far the largest market for sustainable finance products and currently influencing regulations worldwide.

2.3.1 The UN Global Compact

Through the UN Global Compact, the UN provides a voluntary scheme for ESG reporting. By becoming a participant or signatory, companies commit to following ten principles laid out by the UN Global Compact [1]:

1. Businesses should support and respect the protection of internationally proclaimed human rights.
2. Businesses should make sure that they are not complicit in human rights abuses.
3. Businesses should uphold the freedom of association and the effective recognition of the right to collective bargaining.
4. Businesses should uphold the elimination of all forms of forced and compulsory labor.
5. Businesses should uphold the effective abolition of child labor.
6. Businesses should uphold the elimination of discrimination in respect of employment and occupation.
7. Businesses should support a precautionary approach to environmental challenges.
8. Businesses should undertake initiatives to promote greater environmental responsibility.
9. Businesses should encourage the development and diffusion of environmentally friendly technologies.

Table 2.2 Core metrics for ESG reporting provided by the ATHEX (see Footnote 2)

Metrics	Category	Description	Unit of measure
Environmental metrics	Scope 1 emissions	Total amount of direct emissions	Tons CO_2 equivalent
	Scope 2 emissions	Total amount of indirect emissions	Tons CO_2 equivalent
	Energy consumption within the organisation	Total amount of energy consumption (Total energy consumed within the organisation, percentage electricity, and percentage renewable)	MWh, %
Social metrics	Female employees	Percentage of female employees	%
	Female employees in management positions	Percentage of management employees that are women	%
	Turnover rates	Full time employee voluntary and involuntary turnover rate	%
	Employee training	Average hours of training that the organisation's employees have undertaken during the reporting period, by employee seniority	No. of hours
	Human rights policy	Description of human rights policy and fundamental principles	Discussion and analysis
	Collective bargaining agreements	Total number of employees covered by collective bargaining agreements	%
	Supplier assessment	Discussion of supplier screening using ESG criteria	Discussion and analysis
Governance metrics	Sustainability oversight	Discussion on whether the Board of Directors (BoD) provides sustainability oversight at the board committee level or whether sustainability is discussed with management during BoD meetings	Discussion and analysis
	Business ethics policy	Description of business ethics policy and fundamental principles	Discussion and analysis
	Data security policy	Description of data security policy and fundamental principles	Discussion and analysis

10. Businesses should work against corruption in all its forms, including extortion and bribery.

So far, over 12000 companies and other entities have committed to the UN Global Compact worldwide. These participants and signatories must report annually on the progress they make regarding the ten principles outlined above. Adopting ESG measures through the UN Global Compact provides both financial and non-financial benefits to the firms [24]. Points 7–9 are most relevant to material circularity. The depletion of natural resources is deteriorating the environment, especially in the case of fossil fuels. By achieving circularity, societies put a much smaller strain on the environment. As a first step, businesses, especially those involved in the production of goods, should be geared towards circularity and take responsibility for their operations and their supply chain. However, currently available technologies are too limited to achieve circularity while maintaining our standard of living. Thus, innovation toward more environmentally-friendly circular technology is necessary. The other principles aid in boosting the social aspect of circularity.

2.3.2   The EU and EEA

The member states of the European Union, together with the members of the European

Table 2.3 Advanced metrics for ESG reporting provided by the ATHEX (see Footnote 2)

Metrics	Category	Description	Unit of measure
Environmental metrics	Scope 3 emissions	Total amount of other indirect emissions	Tons CO_2 equivalent
	Climate change risks and opportunities	Discussion of climate change-related risks and opportunities that can affect business operations	Discussion and analysis
Social metrics	Stakeholder engagement	Discussion of organisation's main stakeholders and analysis of key stakeholder engagement practices	Discussion and analysis
	Employee training expenditure	Total amount of expenditure on employee training	Euros
	Gender pay gap	Difference between male and female earnings	%
	CEO pay gap	Ratio of CEO to median employee earnings	Ratio
	Sustainable product revenue	Percentage of turnover from sustainable products and services	%
Governance metrics	Business model	Discussion of business model and the creation of value	Discussion and analysis
	Materiality	Description of the materiality assessment process	Discussion and analysis
	ESG targets	Disclosure of short, medium, and long-term performance	Discussion and analysis
	Variable pay	Percentage of executive's salary variable	%
	External assurance	Discussion of external assurance on reported ESG information	Discussion and analysis

Economic Area have the most developed non-financial reporting regulations [1]. The first regulations to create EU-wide reporting standards targeting large undertakings and groups were the 2013 and 2014 directives regarding the disclosure of non-financial and diversity information [25]. These directives have been adopted in national law by all EU and EEA member states as well as in the United Kingdom (which at the time was still a member of the EU). Under the directives [2], large companies are to provide public information on:

1. Environmental matters,
2. Social matters and treatment of employees,
3. Respect for human rights,
4. Anti-corruption and bribery, and
5. Diversity on company boards (in terms of age, gender, educational, and professional background).

The EU directives did leave some room for interpretations which caused some differences between the national laws. Differences are mainly found in which companies have to comply with the reporting regulations, which formats have to be used for the disclosure, and which penalties are given for non-compliance [22]. 64% of EU stock exchanges publish guidelines for ESG reporting for the listed companies in an attempt to standardize ESG reporting, which can be followed by the companies voluntarily [22]. In practice, the voluntary action has resulted in only 17% of the EU's large companies publicly reporting on environmental regulation as per the directives in 2019 [22]. Roughly 88% of the large EU companies that publish environmental governance information address the topic of waste, and 73% address the use of natural resources [22]. The increasing number of reporting requirements and frameworks is leading to an increasing amount of inconsistencies between companies, or even within a company when it comes to reports

Table 2.4 Sector-specific metrics for ESG reporting provided by the ATHEX (see Footnote 2)

Metrics	Category	Description	Unit of measure
Environmental metrics	Emission strategy	Discussion of long- and short-term strategies in relation to the management, mitigation, performance targets of its emissions	Discussion and analysis
	Air pollutant emissions	Total amount of: NO_x (excluding N_2O), SO_X, Volatile organic compounds, and particulate matter (PM10)	kg
	Water consumption	Total amount of: water withdrawn (by source), water consumed, and percentage recycled	m^3, %
	Water management	Discussion of water management risks and the respective mitigation measures taken	Discussion and analysis
	Waste management	Total amount of hazardous and non-hazardous waste generated and percentage of waste by type of treatment (Recycled, treated, and landfilled)	Tons, %
	Environmental impact of packaging	Total amount of hazardous and non-hazardous waste generated and percentage of waste by type of treatment (Recycled, treated, and landfilled)	Discussion and analysis
	Backlog cancellations	Total number of backlog cancellations associated with community or ecological impacts	Number
	Critical materials	Description of management approach in relation to the use of critical materials	Discussion and analysis
	Chemicals in products	Discussion of processes to assess and manage risks and/or hazards associated with chemicals in products	Discussion and analysis
Social metrics	Product recalls	Total number of recalls issued	Number
	Customer privacy	Number of users whose information is used for secondary purposes	Number
	Legal requests of user data	Number of: Law enforcement requests for user information, users whose information whase requested, and percentage resulting in disclosure	Number, %
	Labour law violations	Total amount of monetary losses because of legal proceedings associated with labour law violations	Euros
	Data security and privacy fines	Total amount of monetary losses because of legal proceedings associated with data security and privacy	Euros
	Health and safety performance	Total recordable: Number of injuries, number of fatalities, accident frequency rate, and accident severity rate	Number
	Marketing practices	Description of approach in providing transparent product and service information including marketing and labelling practices	Discussion and analysis
	Customer satisfaction	Disclosure of customer satisfaction survey results	Discussion and analysis
	Customer grievance mechanism	Description of key operations and procedures of customer grievance mechanism	Discussion and analysis
	ESG integration in business activity	Description of approach to incorporation of ESG factors in business activity	Discussion and analysis
Governance metrics	Business ethics violations	Total amount of monetary losses because of business ethics violations	Euros
	Whistle-blower policy	Description of whistle-blower policies and procedures	Discussion and analysis

from different years, on what is reported, the used definitions, and the scope. The regulations and frameworks are often hard to comply with for SMEs in particular [22]. While SMEs are not directly obligated to report on ESG publicly in the EU and EEA countries, large companies increasingly ask for ESG reports from SMEs in their supply chain as they need the information for their reports.

Sustainable finance is considered to play a key role in realizing the EU's vision for a sustainable future as outlined in the European Green Deal [1] and to comply with international commitments such as the Paris agreement [26]. A separate action plan on sustainable finance has also been released [2] as well as a strategy for financing the transition to a sustainable economy that relies on sustainable finance [3]. The European Commission is currently working on several directives, regulations, and other instruments to further develop ESG reporting requirements and standards.

2.3.2.1 Corporate Sustainability Reporting Directive

In April 2021, the European Commission proposed a new Corporate Sustainability Reporting Directive [23]. The proposed directive would task the EFRAG with developing the draft of standards for ESG reporting. EFRAG will publish their first version of the draft by mid-2022 and will consider external expert advice for the same. Besides, the European Commission will task the Member States Expert Group on Sustainable Finance as well as the European Securities and Markets Authority with providing consultations and feedback to the commission. The European Commission will also ask the European Banking Authority, the European Insurance and Occupational Pensions Authority, the European Environment Agency, the European Union Agency for Fundamental Rights, the European Central Bank, the Committee of European Auditing Oversight Bodies, and the Platform on Sustainable Finance to provide inputs to improve the draft. One of the missions given to EFRAG is to comply with existing international (voluntary) standards that have seen broad adoption.

Furthermore, the proposed directive aims to build on the EU Taxonomy regulation. The proposed directive—

1. will make ESG reporting mandatory for all large companies as well as all companies listed on regulated markets (except listed micro-enterprises),
2. compel the audit (assurance) of reported information,
3. will extend the reporting requirements with more detailed provisions that include following EU sustainability reporting standards, and
4. requires companies to digitally tag the reported information, so it is machine-readable and compatible with the European single access point which is developed for the capital markets union action plan[3] [1].

The new directive is being developed as the old directives on the disclosure of non-financial and diversity information were deemed insufficient [23]. Company reports often do not contain all the information investors and other stakeholders increasingly look for. Furthermore, it was difficult to compare information between companies as there were no enforced standards. Investors in the EU and EEA need accurate information on the non-financial performance of companies to comply with the EU regulation on sustainability-related disclosures [27].

The new directive puts a special focus on SMEs. A separate reporting standard is proposed for SMEs so that they do not lose too many resources on creating ESG reports but can still use standardized ESG reports to enable better access to the sustainable finance market. Listed SMEs will be required to follow this simplified standard, non-listed SMEs can follow the standard voluntarily according to the proposal.

2.3.2.2 The EU Taxonomy Regulation

In 2020, the EU approved the EU taxonomy regulation, allowing for delegated acts that establish a taxonomy to facilitate sustainable investments

---

[3] Communication from the European Commission (Retrieved on 04-12-2012) https://eur-lex.europa.eu/legal-content/EN/TXT/?uri=COM:2020:590:FIN.

[28]. In July 2021, a delegated act of the EU Taxonomy was released that obligates firms to publish specified key performance indicators on their turnover, capital, and operation cost linked to environmentally sustainable economic activities, with the act further specifying what counts as such activity [1]. The taxonomy is compliant with the existing non-financial reporting directive as well as the proposed Corporate Sustainability Reporting Directive, it applies to the companies specified by those directives. The delegated act, through defining sustainable activities, harmonizes ESG reporting and helps investors to decide on what is a sustainable investment. Companies are aided in being able to better communicate their non-financial performance and try to outcompete competition as the equal reporting standards create a fairer market. The focus of this first delegated act was mainly on climate mitigation and climate adaptation activities.

In 2022, the European Commission is expected to launch a delegated act for the EU Taxonomy Regulation that defines what products/processes qualify as CE activities as well as for the sustainable use and protection of water and marine resources, protection and restoration of biodiversity and ecosystems, pollution prevention and control, as well as providing a few additional specifications for the climate change mitigation and adaption activities. In August 2021, a draft report was published by the EU's Platform on Sustainable Finance with their work to date [1, 2]. The sectors covered in the draft are:

1. Agriculture, Forestry, and Fishing,
2. Construction and Buildings,
3. Information Communication Technology, and Emergency Services,
4. Mining and Processing,
5. Manufacturing,
6. Energy,
7. Transport,
8. Restoration and Remediation, and Tourism, and
9. Water supply, Sewerage, and Waste Management.

For each of the sectors, technical specifications are presented to define what qualifies as circular activities. This provides a clear understanding of what is considered part of the CE, harmonizes reporting on these activities, and better enables sustainable finance to fund these activities; ultimately speeding up the transition to a CE.

The EU Taxonomy Regulation provides the following definition of the CE–

An economic system whereby the value of products, materials, and other resources in the economy is maintained for as long as possible, enhancing their efficient use in production and consumption, thereby reducing the environmental impact of their use, minimizing waste, and the release of hazardous substances at all stages of their life cycle, including through the application of the waste hierarchy.

The taxonomy regulation also specifies if an activity qualifies as contributing substantially to the transition to a CE. The draft of the delegated act proposes to categorize CE-related activities into one of four categories (listed below) and related issues, based on a yet unpublished report by the EU's Joint Research Centre titled 'Development of the EU Sustainable Finance Taxonomy— A framework for defining substantial contribution for environmental objectives'.

1. Circular design and production
   - Lifetime management
   - Material choices
   - Design for end-of-life management
   - Closed-loop production processes.
2. Circular use
   - Life extension
   - Use intensification.
3. Circular value recovery
   - Preparation for re-use (e.g., resale, repair, remanufacturing, etc.)
   - Recycling
   - Recovery.
4. Circular support
   - Enablers of solutions to the above issues (e.g., digital tools, consulting, predictive maintenance, etc.)
   - Enablers at the interface between activities (e.g., waste trade, charity shops, IS, infrastructure, etc.).

Companies can list their activities if it complies with the provided screening criteria by the EU through an upcoming delegated act [1]. Though, these activities should not harm any of the other objectives set out in the EU's delegated acts of the taxonomy. They also have to comply with a set of basic social safeguards. If these conditions are met, then a company can, and should, list that activity as sustainable in their non-financial performance reports and they are eligible for sustainable finance.

2.3.3 The United States

While the United States has one of the most advanced financial reporting systems, the country is falling behind in the area of non-financial reporting [29]. The US SEC is a government agency which is responsible for protecting investors and should ensure that they are provided with material information to make informed investment decisions. The SEC has issued some mandatory disclosure on corporate governance, such as pay ratio disclosure. However, when it comes to voluntary disclosure, the US has a relatively large number of provisions to report on. Over half of the voluntary provisions are on environmental governance. The Chair of the SEC appointed in 2021 has directed staff to consider mandatory provisions on data disclosure if sustainability claims are made [1].

2.3.4 Canada

In Canada, compared to the US, there are more mandatory provisions to report non-financial information. However, the majority of these provisions go to requesting authorities and not to the public mainstream (annual) reports [29]. Information that has to be publicly disclosed is often corporate governance-related. Reporting to authorities in most cases concerns environmental provisions, such as the use of pollutants, emissions, and environmental incidents.

2.3.5 Japan

Japan (with its non-financial reporting regulations) is not among the best-performing countries, but is quickly improving [30]. Currently, Japan has a 'comply or explain' model for environmental disclosure. However, the country's financial regulator, the FSA, has created a working group with the mission to create mandatory provisions on ESG disclosure with more options to enforce it [2, 3]. The FSA is also working on a certification for ESG funds with further provisions to avoid greenwashing. Voluntary ESG reporting amongst listed firms in Japan has proven to correlate with higher abnormal returns on the stock market [31].

2.3.6 China

China's mandatory non-financial reporting provisions have so far focused mainly on corporate governance. However, since 2020, the CSRC is drafting new rules for listed companies in the area of environmental and social governance while also streamlining corporate governance reporting. At least two consultation rounds on the topic have been concluded in 2021 [1, 2]. A mix of mandatory and voluntary provisions is proposed in the draft. Currently, the proposals contain low penalties for breaches and are thus, not seen as a strong motivator. However, it is expected that the penalties will be increased in the coming years. China aims to have a nationwide environmental disclosure system by 2025 that will be a driving force in achieving sustainable growth and reaching emission targets. China's goals for introducing non-financial reporting regulations seem inspired by the large growth of China's ESG debt market in recent years. Sustainable bond volumes in the Asia-Pacific region have shown over 200% year-on-year growth recently. This is in part due to recent green bond innovations in China [3]. The lack of non-financial reporting regulation has scared off major investors from the Chinese ESG debt market [4]. Companies with good ESG profiles and good ESG reporting have shown both higher returns and reduced credit spreads in China [32, 33].

Fig. 2.2 Sustainable assets under management in billions of USD, split by region. Adapted from UNCTAD [5]

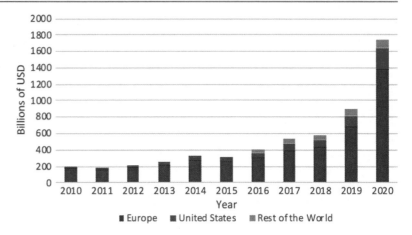

Fig. 2.3 Size of sustainable bonds market, split by type of bond. Adapted from NN Investment partners and UNCTAD [5]

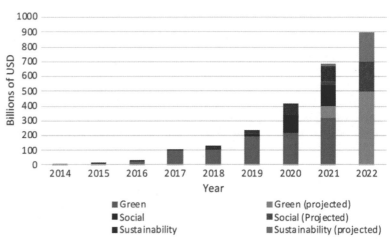

2.4 The Importance of ESG Reporting for Sustainable Finance and Bonds Markets

As previously mentioned in the introduction of this chapter, ESG reporting is beneficial both financially and non-financially to companies and other organizations. Good ESG performance is known to lead to better access to financial markets [3, 4, 6]. ESG initiatives are better received by the market than other sustainability initiatives [7]. ESG policy adoption also increases innovation, for example by providing better access to finance for research and development [9]. ESG metrics also enforce each other through complex inter-dependencies [15]. Besides direct performance, ESG reporting is also increasingly required for public procurement/tenders, to attract subsidies,

and to get access to the rapidly growing market of sustainable finance.

The UNCTAD estimated that the market for sustainability-dedicated investments for 2020 amounted to 3.2 trillion USD, an 80% rise over 2019 [5]. Figure 2.2 shows the historic growth of sustainability funds and assets from 2010 to 2020 (numbers for 2020 are as of June 30). The largest market by far, for sustainable financial products, has been Europe for the last decade. Other regions, especially the US, have recently shown a large growth in the ESG market as well and might gain on soon as the sustainable finance market matures [5].

Sustainable bonds are much like regular bonds, but with the additional commitment from the issuer to use the money towards sustainable investments. Sustainable bonds, totaling 1.5

trillion USD, represent only 1.26% of the global bond market. However, this financial product shows an average annual growth rate of 67% [5]. The UNCTAD expects this market to still be in its early growth stage, expecting the market to grow to 6 trillion USD (5% of the global bonds market). This growth is facilitated by the increased adoption of 'green exchange markets' by stock exchanges. 37 stock exchanges offered such markets as of July 2021, with the first one opening in 2011. Sustainable bonds can be categorized into green bonds, social bonds, and mixed-sustainability bonds. Green bonds first emerged in 2014 but since then have become a 300 billion USD market by 2021, as can be seen in Fig. 2.3. The Green bond growth was lower than expected in 2020, which is likely caused by deferred projects because of the onset of the COVID-19 pandemic [5].

Banks play an important role also through the issuance of green loans and 'sustainability-linked loans'. Green loans are specifically used to finance green projects and assets, and sustainability-linked loans are tied to a borrower's ESG-rating, not specifically to how the money is used. Sustainability-linked loans are a relatively new financial product that came into existence in 2018 with the establishment of the Green Loan Principles. Later on, sustainability-linked loan principles were published by the Loan Market Association, an interested group of banks [33]. Sustainability-linked loans saw a meteoric rise in 2018 and 2019 as can be seen in Fig. 2.4. Such loans are primarily issued in the EMEA region, but the North America and Asia-Pacific regions have been catching up lately.

The adoption of ESG reporting, because of a lack of regulated standards and requirements, is mainly driven by stock exchanges trying to push non-financial reporting through self-regulation. The number of stock exchanges engaged in promoting ESG has risen considerably in the last decade [33]. ESG reporting and other non-financial regulations are increasingly attracting the interest of government regulators in recent years, pushed by a need to transition to more sustainable economies and increasingly realizing the financial benefits of sustainability reporting.

The seeming concentration of sustainable finance in Europe, and the fact that European regulation is, or soon will, surpass the depth of regulation in other markets, creates the risk that the EU will set a global standard for sustainable finance regulation. Within the regulatory competition, a so-called 'race-to-the-top' means that companies try to make their products comply with the toughest regulations, even in markets where such regulations are not in effect. An example of this would be the Brussels Effect [34–36] caused due to the European regulation. The risk here is that the regulation coming from the EU is often designed with a very Euro-centric mindset. Indigenous methods in developing countries that should be considered circular, might not be covered by the European taxonomy on circularity and thus not eligible for sustainable finance and/or properly covered in ESG reporting. This might lead to underdeveloped sustainable finance markets in those countries, and a slower transition towards the CE.

2.5 Lessons Learnt from ESG Metrics and Reporting

ESG reporting is gaining traction, evidenced by increased adoption by corporations and regulation of reporting standards. ESG reporting serves a number of purposes. Mainly, it indicates the sustainability performance of a firm to a wide range of stakeholders and provides access to a growing market of sustainable financial products, from loans to bonds and investments (a market that shows a strong annual double-digit growth and is expected to reach a market cap of 6 trillion USD globally by 2025 according to UNCTAD). Subsidies and public procurement tenders are increasingly relying on ESG reports as well. Another interesting development is the emergence of ESG rating agencies and the closely linked sustainability-linked loans.

ESG reporting provides valuable information not only for investors but for many other (external) stakeholders such as policymakers. When a significant part of a local/regional/national/supranational economy

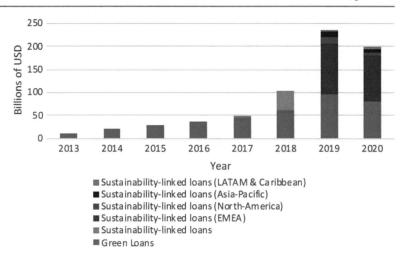

Fig. 2.4 Size of green loan and sustainability-linked loan markets, split by region for 2019 and 2020. Adapted from UNCTAD [5]

is covered in ESG reports, that information can be used effectively by policymakers to tailor their work. To be able to aggregate the data from those ESG reports, it is important that many firms/organizations publish them and that they follow the same reporting standards. Currently, many countries have systems in place to aggregate data on CO_2 emissions, but this is only done for a few other ESG metrics. A problem with aggregating ESG reports is that reports typically present consolidated data for the entire company/organization spread over different parts of the world. This consolidation can make it hard to localize the impacts of ESG issues, making it less valuable to policymakers. The next step in ESG reporting could very well be the disclosure of localized ESG reports, that form a subset of the consolidated companywide ESG report, to better inform employees and local policymakers about specific conditions in a specific area of interest.

Besides the standardization and mandatory enforcement of ESG reporting, the European Union plans to standardize the digital reporting of both financial and non-financial (including ESG reports) information to better aggregate such data for macro analysis. These efforts are part of the creation of a single capital markets union [1–4]. In addition, a single point to access this digitally disclosed information is also created for the benefit of investors, analysts, asset managers, consumers, NGOs, data vendors, credit risk assessment entities, and banks amongst others [6, 7].

A big step to consider in the coming year is the standardization of ESG reports. Right now, ESG reporting suffers from a lack of consistency. The results of different firms are hard to compare with each other, and even within organizations, the results from one year to the next can be hard to compare. Standardization has so far been driven mainly through self-regulation, often organized by stock exchanges. Several NGOs have provided principles and standards that are being increasingly adopted globally. Regulators are still working on enforcing reporting standards, the first comprehensive regulation in this area is expected to take effect in Europe in 2023.

2.5.1 A Call for Clear Circularity Assessment Within ESG Reporting

Current voluntary ESG reporting standards have metrics relevant to the CE, such as the metrics related to the use of critical raw materials and water. However, many aspects of the CE are yet to be properly captured in current best practice ESG reporting metrics. Partially, this is due to the lack of a clear universal definition of what accounts for sustainable circular practice and what doesn't. Therefore, there is a need for the integration of ISO standards for circularity assessment with standardized ESG reporting in the future. Since the 2015 Paris agreement, countries have agreed to create binding taxonomies on what counts as sustainable and what doesn't. In coherence with this agreement, the European Union will release a taxonomy specific to CE practices in 2022. This must be tied to ESG reports to enable clear comparisons between organizations on how circular they are.

Sustainable finance forms a considerable driver for firms to become more sustainable. With specific taxonomies defining which circularity-related practices are considered sustainable, this funding will also become available for companies transitioning towards a circular business model. With the expected growth of the sustainable finance sector, investors- both institutional and individual, will increasingly provide a push to enhance circularity. For this reason, it is also crucial that a company's circularity be reflected in its ESG reporting, allowing investors to make better decisions and creating a further driver for companies to become more circular. ESG regulation and sustainability taxonomies (including circularity) thus provide a powerful tool for governments to steer the market towards investments in the CE.

In Part II of this book on measuring circularity, we provide the indicators currently used for assessing various systemic levels from macro to the nano, most of which have relevance to comprehensive ESG reporting.

References

1. Zubeltzu-Jaka E, Andicoechea-Arondo L, Alvarez Etxeberria I (2018) Corporate social responsibility and corporate governance and corporate financial performance: bridging concepts for a more ethical business model. Bus Strateg & Devel 1(3):214–222
2. Thostrup Jagd J, Krylova T et al (2018) Reporting on the sustainable development goals- [english] - a survey of reporting indicators
3. Cheng B, Ioannou I, Serafeim G (2014) Corporate social responsibility and access to finance. Strateg Manag J 35(1):1–23
4. Nollet J, Filis G, Mitrokostas E (2016) Corporate social responsibility and financial performance: non-linear and disaggregated approach. Econ Modell 52:400–407
5. United Nations Conference on Trade and Development, Team led by James X Zhan (2021) World invest-
ment report 2021 – Investing in sustainable recovery
6. Cheng MM, Green WJ, Ko JCW (2015) The impact of strategic relevance and assurance of sustainability indicators on investors' decisions. Aud: J Pract & Theory 34(1):131–162
7. Lo KY, Kwan CL (2017) The effect of environmental, social, governance and sustainability initiatives on stock value-Examining market response to initiatives undertaken by listed companies. Corpor Soc Responsib Environ Manag 24(6):606–619
8. Widyawati L (2020) A systematic literature review of socially responsible investment and environmental social governance metrics. Bus Strat Environ 29(2):619–637
9. Broadstock DC, Matousek R, Meyer M et al (2020) Does corporate social responsibility impact firms' innovation capacity? The indirect link between environmental & social governance implementation and innovation performance. J Bus Res 119:99–110
10. Minutolo M, Kristjanpoller W, Stakeley J (2019) Exploring environmental, social, and governance disclosure effects on the S&P 500 financial performance. Bus Strat Environ 28:1083–1095
11. Tamimi N, Sebastianelli R (2017) Transparency among s&p 500 companies: an analysis of esg disclosure scores. Management decision, pp 1660–1680
12. Ruan L, Liu H (2021) Environmental, social, governance activities and firm performance: evidence from China. Sustainability 13(767):1–16
13. Sultana S, Zainal D, Zulkifli N (2017) The influence of environmental, social and governance (ESG) on investment decisions: the Bangladesh perspective. Pertanika J Soc Sci Humanit 25:155–173
14. Yoo S, Managi S (2022) Disclosure or action: evaluating ESG behavior towards financial performance. Finan Res Lett 44(102):108
15. Jitmaneeroj B (2016) Reform priorities for corporate sustainability: environmental, social, governance, or economic performance? Manag Dec 54(6):1497–1521
16. Whitelock VG (2015) Environmental social governance management: a theoretical perspective for the role of disclosure in the supply chain. Int J Bus Inf Syst 6 18(4):390–405
17. Rodríguez-Fernández M, Sánchez-Teba EM, López-Toro AA et al (2019) Influence of ESGC indicators on financial performance of listed travel and leisure companies. Sustainability 11(5529):1–20
18. Paolone F, Cucari N, Wu J et al (2021) How do ESG pillars impact firms' marketing performance? A configurational analysis in the pharmaceutical sector. J Bus & Ind Market 1–13
19. Cucari N, Esposito de Falco S, Orlando B (2018) Diversity of board of directors and environmental social governance: evidence from Italian listed companies. Corp Soc Respons Environ Manag 25(3):250–266
20. Nicolò G, Zampone G, Sannino G et al (2021) Sustainable corporate governance and non-financial disclosure in Europe: does the gender diversity matter? J Appl Acc Res 23(1):227–249

21. Lagasio V, Cucari N (2019) Corporate governance and environmental social governance disclosure: a meta-analytical review. Corp Soc Respons Environ Manag 26(4):701–711
22. European Financial Reporting Advisory Group (2021) Current non-financial reporting formats and practices
23. European Commission (2021) Proposal for a directive of the European parliament and of the council amending Directive 2013/34/EU, Directive 2004/109/EC, Directive 2006/43/EC and Regulation (EU) No 537/2014, as regards corporate sustainability reporting
24. Ortas E, Álvarez I, Garayar A (2015) The environmental, social, governance, and financial performance effects on companies that adopt the United Nations Global Compact. Sustainability 7(2):1932–1956
25. European Parliament and Council of the European Union (2014) Directive 2014/95/EU of the European Parliament and of the council of 22 October 2014 amending Directive 2013/34/EU as regards disclosure of non-financial and diversity information by certain large undertakings and groups
26. European Parliament and Council of the European Union (2021) Regulation (EU) 2021/1119 of the European parliament and of the council of 30 June 2021 establishing the framework for achieving climate neutrality and amending Regulations (EC) No 401/2009 and (EU) 2018/1999 ('European Climate Law')
27. European Parliament and Council of the European Union (2019) Regulation (EU) 2019/2088 of the european parliament and of the council of 27 November 2019 on sustainability-related disclosures in the financial services sector
28. European Parliament and Council of the European Union (2020) Regulation (EU) 2020/852 of the European parliament and of the council of 18 June 2020 on the establishment of a framework to facilitate sustainable investment, and amending Regulation (EU) 2019/2088
29. World Business Council for Sustainable Development (2018) Corporate reporting in the United States and Canada
30. Singhania M, Saini N (2021) Quantification of ESG regulations: a cross-country benchmarking analysis. Vision pp 1–9
31. Nakajima Y, Inaba Y (2021) Stock market reactions to voluntary integrated reporting. J Fin Report Acc 1–26
32. Zhang X, Zhao X, Qu L (2021) Do green policies catalyze green investment? Evidence from ESG investing developments in China. Econ Lett 207:1–3
33. Yang Y, Du Z, Zhang Z et al (2021) Does ESG Disclosure Affect Corporate-Bond Credit Spreads? Evidence from China. Sustainability 13(15):1–15
34. Bradford A (2015) Exporting standards: the externalization of the EU's regulatory power via markets. Int Rev Law Econ 42:158–173
35. Bradford A (2012) The Brussels effect. Northwestern Univ Law Rev 107:1
36. Wu HH (2020) Territorial extension of the EU environmental law and its impacts on emerging industrial economies: a Taiwan case. China WTO Rev 6(2):325–350

Part II
Measuring Circularity

... 'circular' approaches–to the city, the economy, design–extend well beyond just limiting environmental impacts. They take on a more systemic, cyclical view of how physical and biological processes, together with human interactions, give rise to sustainable living environments–forming a complete self-sustaining 'ecosystem', like a closed circle.

—Michiel Schwarz

in A Sustainist Lexicon: Seven Entries to Recast the Future
—Rethinking Design and Heritage

Initiator and Director, Sustainism Lab
Amsterdam, North Holland, Netherlands
Co-creator, The *Sustainism* manifesto

Circularity at Macro Level: The Urban and National Perspectives

Patrizia Ghisellini, Sven Kevin van Langen, Rashmi Anoop Patil and Seeram Ramakrishna

Abstract

The CE is mainly conceived as a response to multiple global environmental and social challenges, including climate change and resource scarcity. This chapter first presents an overview of what entails the adoption of CE in cities. Later, CE in other macro-level systems (such as nations and wider areas) and the necessity for circularity assessment during the transition towards CE are discussed. From a global perspective, cities play a crucial role in contributing to tackle the climate challenges given the high consumption of energy and materials. The discussion elaborates on assessment frameworks and indicators for monitoring and evaluating the progress towards the CE at the city and regional/national/supra-national (e.g. The EU)/global levels. Case studies of Rotterdam and Paris are provided to show how cities have designed their circular plans by analyzing their strategies, tools, and performance indicators. The monitoring framework implemented by the EU, to monitor the transition to CE in its member states is also analyzed. The chapter concludes by pointing out the importance of promoting preventive measures to enhance the circularity and broader assessment framework that captures the social dimension of the CE.

P. Ghisellini(✉)
Department of Engineering, Parthenope University of Naples, 80143 Naples, Italy
e-mail: patrizia.ghisellini@gmail.com

S. K. van Langen
UNESCO Chair in Environment, Resources and Sustainable Development (International Ph.D. Programme), Department of Science and Technology, Parthenope University of Naples, 80143 Naples, Italy

Olympia Electronics, Thessaloniki, Greece

R. A. Patil · S. Ramakrishna(✉)
The Circular Economy Task Force, National University of Singapore, Singapore 117575, Singapore
e-mail: seeram@nus.edu.sg

S. Ramakrishna
Department of Mechanical Engineering, National University of Singapore, Singapore 117575, Singapore

Centre for Nanotechnology and Sustainability (NUSCNS), 2 Engineering Drive 3, Singapore 117576, Singapore

Keywords

Macro-level circularity assessment · Environmental sustainability · Urban circularity · Circular cities · National indicators

© The Author(s) 2023
R. A. Patil and S. Ramakrishna (eds.), *Circularity Assessment: Macro to Nano*,
https://doi.org/10.1007/978-981-19-9700-6_3

3.1   Introduction

The interest in Circular Economy (CE) as a concept and model of economic system is growing faster than ever all over the world [1]. The CE is considered as one of the pillars of response to pressing environmental issues [2, 3] and the declared climate emergency [4] as well as a tool for establishing sustainable development [2, 5]. The CE promotes changes in the whole society and its subsystems (economic, political, cultural, technical-technological). In the economic subsystem, resources (materials, waste, energy) entering into production and consumption activities should be (according to CE) used more efficiently by narrowing, slowing and closing resources loops [6]. The CE advocates the principles of:

- designing out waste and pollution;
- prolonging the use/value of materials and products via the most suitable strategies;
- regenerating the natural (eco) systems, by creating a better balance with human activities and their conurbations.[1]

In recent years, the EU, China and Japan being pioneers in adopting circular strategies [7, 8], have continued to guide the process of transition towards CE, and many other countries are beginning to follow in their footsteps.[2,3] Programs, policies and legislation have been adopted to pursue the CE goal and accelerate the transition [7–12]. While cities such as Amsterdam, Rotterdam, Brussels, London, and Paris are adopting circular strategies city-wide, other countries (such as China) are aiming more directly at sector integration (such as industrial parks, civic waste) and mapping circularity[4] according to their development pattern [7]. The authors' opinion is that the process of gradual transition from the linear to the CE at all macro levels (cities, regions, nations, and supra-national institutions) has begun. Therefore, monitoring and evaluating the progress towards CE and the effects of circular strategies on a large/macro scale (cities and countries) is necessary. Several assessment tools and indicators have evolved to be the most common methods of choice, for monitoring the CE implementations at the macro level [13–22] such as-

- Material Flow Accounting (MFA) [23, 24],
- Input Output Analysis, and
- LCA [13, 14, 17, 18, 25, 26].

In some cases, these methods are combined by applying e.g., MFA with LCA as well as with Life Cycle Costing [27] and Social LCA [25]. Additionally, MFA is suggested in combination with EMergy Accounting (EMA) [28, 29]. The latter is also proposed individually, to monitor and evaluate the CE implementations for their effects on the wider dimension of the Biosphere, accounting for the total energy required, directly or indirectly, to make a given product or support a flow [30].

In this chapter we mainly focus on indicators for monitoring and evaluating the transition to CE at the macro level, that is, creating an overview of how well the principles of CE (reduce, repair, reuse, recover, remanufacturing, and recycle) are implemented in the different types of macro-systems. The discussion begins with the transition to CE within cities, highlighting the importance of cities in a global society—both as cradles of opportunities and positive change, as well as centers of critical challenges for more sustainable development. Then, a detailed discussion on indicators for measuring CE in cities is presented, supported by two case studies of circular cities: Rotterdam (Netherlands) and Paris (France). An overview of indicators developed by the EU for monitoring CE in the whole EU, as well as within each of its member states is provided. The essential elements useful for addressing CE within macro-level systems (cities, regions, nations, economic blocks) are highlighted in the conclusion.

---

[1] Ellen Mac Arthur Foundation (2017) Circular economy in cities https://www.ellenmacarthurfoundation.org/our-work/activities/circular-economy-in-cities.

[2] https://www.innovatorsmag.com/the-leading-circular-economy-nations/.

[3] https://coolerearth.cimb.com/articles/circular-economy-clubs-circular-cities-week-2020-cec-asean-uk-leaders-coming-together.html.

[4] https://www.c40.org/researches/municipality-led-circular-economy.

3.2   Transition to Circular Cities

Cities can be defined as permanent and densely populated spatial systems with administrative defined boundaries where the population is mainly employed in non-agricultural activities [31]. However, in recent years, there are many forms of urban agriculture developing in cities [32–35]. According to the UN statistics, currently, about 50% of the global population live in cities and this figure is expected to double by 2050.[5] Over the years, the concentration of economic and industrial activities in the cities or their proximities has been a catalyst to high migration of people from rural to urban areas attracted by job and wealth opportunities. Currently, cities and metropolitan areas contribute to generating about 60% of global gross domestic product (GDP).[6]

Cities are confronted with many challenges due to this high concentration of human and economic activities. The population clustering adversely affects the urban environment, causing poor air quality, water pollution and other environmental issues[7,8] [36]. Moreover, the high consumption of non-renewable energy (a few forms of fossil fuels in particular) in cities generates significant quantities of greenhouse gases (GHGs) making cities one of the main contributors to climate change [37]. According to the IPCC, climate change is mainly an energy challenge as energy use accounts for over two-thirds of GHGs that cause the alteration of the natural "greenhouse effect".[9] Even though the transition to cleaner energy is on course worldwide, it has not yet reached the level required to mitigate climate change and the ongoing worsening of that phenomenon. Therefore, a greater acceleration of overall actions towards wider use of clean energy is critically required. By implementing CE principles, cities have the chance to provide a significant contribution to the energy transition in transport and other industrial systems and better resource utilization in both production and consumption activities. Cities can lead to regenerating the natural systems in the urban environments and the global environment.[10]

A possible definition of a circular city can be: "A circular city is a city that practices CE principles to close resource loops, in partnership with the city's stakeholders (citizens, community, business, and knowledge stakeholders), to realize its vision of a future-proof city" [38]. This definition focuses both on the capability of cities to "closing the loop" (represented in Fig. 3.1) and managing the implementation of reverse logistics for the reuse, repair and remanufacturing of products. It also focuses on the features of CE as a participative process, promoting social innovation and responsibility of all the stakeholders in the city [38]. Although the transitioning to CE can be with different goals (such as sustainable development, economic benefits or adhering to the regulations), the key areas for intervention and tools for achieving circularity are common [11]. A strong cooperation among all the stakeholders is the key factor within the shared vision of sustainability and circularity in the city. The decisions in this view have a local dimension but their effects can also be global[11] [39]!

So far, the EU, including the European Investment Bank,[12] and international institutions such as the UN (within the UN's Global Initiatives

---

[5] United Nations (2020) Sustainable cities, why they matter? https://www.un.org/sustainabledevelopment/wp-content/uploads/2019/07/11_Why-It-Matters-2020.pdf.

[6] https://www.un.org/sustainabledevelopment/cities/.

[7] United Nations, 2019b https://news.un.org/en/story/2019/09/1046662.

[8] National Geographic, Nunez, (2019) https://www.nationalgeographic.com/environment/global-warming/greenhouse-gases/.

[9] IPCC, (2020) https://www.ipcc.ch/2020/07/31/energy-climate-challenge/.

[10] Ellen Mac Arthur Foundation (2017) Cities in the circular economy: an initial exploration https://www.ellenmacarthurfoundation.org/assets/downloads/publications/Cities-in-the-CE_An-Initial-Exploration.pdf.

[11] EU, 2019, Urban Agenda for the EU. Indicators for circular economy (CE) transition in cities—Issues and mapping paper (Version 4) https://ec.europa.eu/futurium/en/system/files/ged/urban_agenda_partnership_on_circular_economy_-_indicators_for_ce_transition_-_issupaper_0.pdf.

[12] European Investment Bank, 2018. The 15 circular steps for cities https://www.eib.org/attachments/thematic/circular_economy_15_steps_for_cities_en.pdf.

Fig. 3.1 Schematic representation of a circular city with various loops representing the circularity of water, energy, nutrients, and reusable materials. Adapted from [40]. Copyright, 2020 Springer Nature

for Resource Efficient Cities[13]) and the Organisation for Economic Co-operation and Development (OECD)[14] have developed many initiatives for supporting the transition towards sustainability and circularity in cities, including guidance for step-by-step implementation of the CE (adapted from the European Investment Bank, 2018 (see Footnote 12)) as outlined below.

- **Plan** (see Footnote 12)
  1. Characterize and analyze local context and resource flows, and identify idle assets
  2. conceptualize options and prioritize among sectors with circular potential
  3. craft a circular vision and strategy with clear circular goals and targets.
- **Act** (see Footnote 12)
  4. Close loops by connecting waste/residue/ water/heat generation with off-takers

13 United Nations Global Initiative for Resource Efficient Cities https:// resourceefficientcities.org/.

14 OECD, The Circular Economy in Cities and Regions http://www. oecd.org/cfe/regionaldevelopment/circular-economy-cities.htm.

5. consider options for extending use and life of idle assets and products
6. construct and procure circular buildings, energy and mobility systems
7. conduct hlcircular experimentation—address urban problems with circular solutions
8. catalyse circular developments through regulation, incentives and financing;
9. create markets and demand for circular products and services—be a launching customer
10. capitalize on new ICT tools supporting circular business models.
- **Mobilise/Monitor** (see Footnote 12)
11. Coach and educate citizens, businesses, civil society and media
12. confront and challenge linear inertia, stressing linear risks/highlighting circular opportunities
13. connect and facilitate cooperation among circular stakeholders
14. contact and learn from circular pioneers and champions
15. communicate on circular progress based on monitoring.

Indicator frameworks for monitoring urban circularity have been proposed by the above-mentioned institutions. The EU Indicator framework for CE is intended for both the national and regional scales and can be applied for building city-level indicators. It consists of indicators for each thematic area such as production and consumption, waste management, secondary raw materials, competitiveness, and innovation[15] (In Sect. 3.6, the EU indicator framework is discussed briefly). Moreover, representative organizations involved in the CE transition (e.g., Ellen Mac Arthur Foundation) are supporting city leaders in embedding the CE in their urban plans and policies.[16]

Many cities worldwide have already adopted plans and strategies for CE [11, 15, 41–43] that also include monitoring frameworks and indicators [15, 44, 45]. Recent literature [11] points out the need for improving the adopted frameworks for circular cities in terms of both programs and indicators. In some cases, cities have CE programs but lack indicators, whereas in other cases the indicators built for the national and regional scale are adapted for the city scale [11]. In this view, the UN emphasizes the importance of measuring the progress towards CE, by focusing not only on conventional indicators, capturing the contribution to economic growth, or on materials and solid waste, but also on people and the qualitative dimension of their wellbeing (noting to give particular attention to minorities and (other) vulnerable subjects of the population).[17]

As discussed in Chap. 1, the UN's proposition for circularity assessment includes social factors along with the environmental and governance counterparts. These three dimensions also appear in the UN's sustainable development initiatives. For example, (i) in the case of mobility, cities should be redesigned, having the people at their center rather than vehicles; (ii) investments should be oriented towards zero-carbon public transport, footpaths, protected paths, and other special routes for walking and biking.[18] In this regard, Paris is planning to become a "15-minute city". Everything a typical citizen needs will never be more than 15 min away from their home by public transport or walking/cycling. Such an approach developed by Moreno aims to reduce the use of fuel-based transport and the resulting CO_2 emissions and other pollutants leading to a better urban air quality.[19]

---

[15] European Commission, 2018. Indicators for sustainable cities. In-depth Report 12, produced for the European Commission DG Environment by the Science Communication Unit, UWE, Bristol http://ec.europa.eu/science-environment-policy.

[16] Ellen Mac Arthur Foundation, 2018. The Circular Economy Opportunity for Urban and Industrial Innovation in China https://www.ellenmacarthurfoundation.org/publications/chinareport.

---

[17] United Nations (2020) Global Indicator Framework for the Sustainable Development Goals and targets of the 2030 Agenda for Sustainable Development https://unstats.un.org/sdgs/indicators/Global%20Indicator%20Framework_A.RES.71.313%20Annex.pdf.

[18] United Nations News (2019) As urbanization grows, cities unveil sustainable development solutions on World Day https://news.un.org/en/story/2019/10/1050291.

[19] World Economic forum, Paris is planning to become a 15-minute city https://www.weforum.org/videos/paris-is-planning-to-become-a-15-minute-city-897c12513b.

The need for relying on broader indicators is in agreement with the targets of the UN's 2030 Agenda for achieving their SDGs [45]. Undoubtedly, the shortcomings related to indicators are a part of the initial stage of the development of these tools. Further efforts by the institutions, cities, and the relevant stakeholders and communities (such as policymakers, universities, citizens, nonprofit organization, and companies) are needed along with learning from experimenting, international experiences, and best practices, as new and improved indicators are developed.[20]

3.3 Indicators for Measuring Urban Circularity

Table 3.1 provides a comprehensive set of indicators regarding transiting to a CE developed by the European circular cities [15] and Chinese cities[10]. The indicators are classified according to the dimension (e.g., environmental, economic, or social) as well as the specific area within each dimension.

From Table 3.1 (elaborated from [10, 15]), it's evident that there are a few common indicators included in the monitoring frameworks of the 13 circular cities as listed below-

- Amount or percentage of recycled material (Tons/year or %/year) used by 4 circular cities (London, Rotterdam, Maribor, Ljubljana),
- CO_2 emissions saved/ GHG emissions saved used by 8 circular cities (London, Glasgow, Marseille, Prague, Malmo, Gothenburg, Kawasaki),
- Amount or percentage of waste avoided used by 9 circular cities (London, Glasgow, Prague, Rotterdam, Antwerp, Paris, Maribor, Malmö, Gothenburg), and
- Number or percentage of new jobs from CE, the share of circular jobs, number of green jobs... used by 12 circular cities (London, Marseille, Amsterdam, Rotterdam, Paris, Glasgow, Maribor, Prague, Ljubljana, Kawasaki).

3.4 Case Study I: Circular Rotterdam

Rotterdam is the second biggest city in the Netherlands, situated in the densely populated Holland region and known for having the largest port in Europe and the 10th largest globally. In 2018, an extensive report on the city's material flows and its potential to transit to a CE was released.[21] They found that in 2018, roughly 10% of the city was circular (the national average being 8.1%) and about 22% of the city's solid waste was recycled (and most of the remaining solid waste is incinerated). Four sectors were identified that have the best potential for reducing the raw material consumption of the city and its population.

First, there is the Agri-Food and Green Flows sector. With 14% of food entering the city being wasted (on average this amounts to 62 kg of food per person per year), food waste is identified as one of the key issues to tackle for reducing the city's environmental footprint and material consumption. Second, there is the construction sector which is responsible for much of the city's waste production, with about 395 kilotons of waste a year, and the logistics in this sector are one of the major contributors to the air pollution within the city. Healthcare is also identified as one of the most critical sectors, though perhaps not the most logical choice for making supply chains circular. The report identifies this sector as having a large potential for reducing material consumption by preventing the need for healthcare/medication (e.g. through promoting healthier lifestyles, reducing pollution, but also by moving healthcare closer to home and reducing frequent hospital visits). Lastly, the consumer goods sector is identified, being one of the larger users of metal (having a similarly sized value to total metal use in construction) to be having a high potential for better value retention as identified from the material flow analysis. These sectors together are responsible for 60% of the city's waste generation and 36.7% of employment at the time of the report.

---

[20] OECD, The Circular Economy in Cities and Regions http://www.oecd.org/cfe/regionaldevelopment/circular-economy-cities.htm.

[21] Rotterdam Circulair, opportunities for new jobs in a zero-waste economy https://rotterdamcirculair.nl/wp-content/uploads/2018/11/GemeenteRotterdam_Report_English_15-11-18.pdf.

Table 3.1 Indicators of the following circular cities or cities in transition to CE: London, Paris, Rotterdam, Maribor, Prague, Amsterdam, Malmö, Glasgow, Marseille, Kawasaki, Antwerp, Brussel, Luibljana, Beijing, Shanghai, Tianjin, and Dalian. *Source* Elaboration from [10, 15]

Dimension of indicators	Area of indicators	Type of indicator	Unit of measure
Environmental	Reverse Cycles (Reuse, Recycling products and materials, raw materials use)	Amount or percentage of recycled material	Tons/year or %/year
		Amount or percentage of products reused	Tons/year or %/year
		Amount or percentage of products recovered	Tons/year or %/year
		Amount of raw materials used in the manufacturing processes	Tons/year
		Average amount of materials retained in the cycle per citizen per year	Kg/year
		Percentage of incoming/outgoing flows	%/year
	Climate Change	Amount of CO_2 emissions	Kg CO_2/year
		Amount of greenhouses gases emissions	Kg CO_2/year
		CO_2 (or CO_2 equivalent) emissions saved (including industrial and urban symbiosis)	Tons/year or Tons CO_2 eq. /year or %/year
		GHG emissions saved (e.g by means of circularity)	Tons/year or Tons CO_2 eq. /year or %/year
		Amount of emissions of NO_x	Tons/year
	Air quality	Amount of emissions of fine dust emissions	Tons/year or PM2.5 $\mu g/m^3$
		Annual average air quality particulate matter	Tons/year or PM2.5 $\mu g/m^3$
		Reduction in embodied carbon (buildings life cycle)	kg CO_2 eq. per kg of product
		CO_2 intensity	Tons/capita
		Air pollution and greenhouse gas emissions associated to transport	Tons/year
	Waste Generation and Management (Reduction, Reused, Recycled)	Average amount of products going to landfill or incineration	Tons/year
		Waste reduction in production of goods-raw material efficiency	Kg of waste per €1000 output
		Amount or percentage of waste separation	%/year or tons/year
		Increase in the clean plastics and drink packaging streams from residual waste	%/year
		Percentage of recycling of the solid waste generated in the city	%/year
		Percentage of recycle of packaging waste; Percentage of recycled municipal waste	%/year
		Difference between tonnes of waste and tonnes of products consumed	Tons of waste/tons of products consumed

(continued)

Table 3.1 (continued)

Dimension of indicators	Area of indicators	Type of indicator	Unit of measure
		Waste diverted via repair, reuse, recovery and upcycling activities (recycling centres, artisans, second-hand goods stores, fab-labs, etc.)	Tons/year
		Municipal waste produced in the city	Tons/year
		Amount of waste generated per capita	Tons/per capita/year
		Amount of waste produced and treated in the city	Tons/year or %/year
		Amount of solid waste reused	Tons/year or %/year
		Amount or percentage of waste avoided	Tons/year or %/year
		Amount of household waste reduced by prevention and reuse	Tons/year or %/year
		Amount of biowastes processed in biogas facilities	Tons/year or %/year
	Energy efficiency, Electricity use, Renewability	Energy consumption per GDP	GJ/GDP
		Energy consumption per unit of industrial value added	GJ/ind.value added
		Use of renewable resources	%/year
		Energy savings per year	%/year
		Absolute (kWh) and relative (%) reduction of yearly electricity consumption	kWh/year or %/year
		Less use of peak power	%/year
		Energy requirement per capita	GJ/person/year
		GDP per energy requirement	€/GJ
		Percentage of renewable or recycled energy use	%/year
		Supply of renewable energy	%
		Renewable energy production on total energy production	MWh/year/total
		Fossil-fuel-free transport sector	%
		Percentage of renewable electricity supply for all municipal operations	%
		Number of families powered by energy produced by wind turbines	N./total
		Electricity consumption per capita	MWh per Capita/year
		Public transit ridership for work and school commutes	%
		Eco-car strategy-Municipal fleet powered by biogas, hydrogen or electricity (including plug-in hybrids)	%/ year
		Percentage of building heating mainly by natural gas	%
		Percentage of building heating mainly by energy from incineration	%
		Percentage of water heating by natural gas	%

(continued)

Table 3.1 (continued)

Dimension of indicators	Area of indicators	Type of indicator	Unit of measure
	Resources use, Resource efficiency and Biodiversity	Primary resources used	%/year or Tons/year
		Raw material consumption	%/year
		Primary raw material demand per capita	Ton/capita
		Virgin resources used	%/year or Tons/year
		Amount of primary resource use avoided	%/year or Tons/year
		Amount of material saving due to the implementation of circular strategies	Tons
		Number of symbioses/synergies connecting businesses (resources exchanged)	N./year
		Reduction of material use due to industrial and urban symbiosis	%
	Water consumption and treatment	Water consumption per unit of industrial value added	m^3/value added
		Water consumption per capita per year	m^3 per capita/year
		Tap water use	%/year
		Liters or cubic metres of household's daily water-consumption	Liters/day
		Rate of municipal wastewater treatment	%
		Rate of wastewater recycling	%
	Mobility	Maritime traffic (amount of maritime throughput)	Tons/year
		Energy consumption referred to transport sector	KWh/year
	Land use	Land covered by a circular platform in the industrial port area	m^2 (Square meters)
		Percentage of the land area used for CE projects implementation	%
		Number of plants involved in circular CE projects	N/year
	Biodiversity	Percentage of biodiversity	%
		Creation of protected green areas	N./year
		Green areas used as stormwater storage	m^2/total surface
		Proportion of green and recreational areas per capita	%
	Buildings	Percentage of annual rainfall absorbed by green roofs	%/year
		Temperature of external facades (decrease for example thanks to green facades)	°C
		Indoor temperatures (decrease for example thanks to green facades)	°C
	Food	Percentage of sustainable food	%

(continued)

Table 3.1 (continued)

Dimension of indicators	Area of indicators	Type of indicator	Unit of measure
Economic and financial	Economic and financial benefits from circular strategies	Money saved by average household due to reducing the amount of products thrown away	€/year
		Financial savings to consumers and businesses adopting more effi	€/year
		cient circular business models	€/year
		Financial savings to public sector bodies through improved procurement practices/waste management	€/year
		Financial savings to consumers from decreased consumption of "new products"	€/year
	Waste Management Costs	Waste management costs	€/year
	Circular and green investments (pilot projects implementing CE, public funds in CE projects)	Budget allocated to support pilot CE projects at the local level	€/year
		Return on investment	€/year
		Public funding in CE projects	€/year
	Environmental costs	Environmental costs (costs of exhaustion, water pollution, CO_2-emissions, eco-toxicity and land use)	€/kg
		Costs related to flood risk	€/kg
	Gross Value Added	Gross value added	€/year
	Economic benefits from selling/purchase of reused products	Total revenue from sale/leasing of reused products	€/year
		Economic savings in purchasing reused products for citizens	€/year
	Economic benefits and revenues from implementing circular strategies	Resource usage: total raw material productivity	GDP/tons of primary input material
		Money granted to businesses or research projects linked to the CE	€/year
		Net added value due to the implementation of circular strategies	€/year
		Value creation from the growth of CE models	€/year
	Economic value of reused products	Average value of products	€
		Value of re-usable or recyclable goods avoiding landfill	€
	Volume of sales, Change GDP, turnover	Volume of sales from the growth of CE models	Amount/year or €/year
		Sales of locally produced goods	Amount/year or €/year
		Revenues through sales thanks to the growth of CE models	€/year
		Change in GDP through circular activities	%

(continued)

Table 3.1 (continued)

Dimension of indicators	Area of indicators	Type of indicator	Unit of measure
		Turnover of organizations working in the CE (including all sectors and types)	€/year
		Global sales related to CE	%/year
		Annual fees related to CE	%/year
		Tenancy turnover	%/year
		Creation of value added and economic growth	€/year
	Economic benefits from urban symbiosis	Economic advantage from industrial symbiosis activities	€/year
		Economic benefits from exchanges of material waste diverted from incineration and landfill	€/year
Social and Cultural	Circular Job creation	Number of new jobs	N./year
		Share of circular jobs (full- or part-time jobs related to one of the seven basic principles of circular employment)	%/year
		Percentage of new jobs related to the CE	%/year
		Number of new jobs from recycling of packaging	N./year
		Number of new green jobs	N./year
	Circular business creation	New business opportunities	N./year
		New businesses that have integrated circularity into their development process	%/year
	Circular training creation (including training employees)	Number of training opportunities related to CE	N./year
		Number of individuals trained through the education measures	N./year
	Unemployment rate	Unemployment rate	%/year
	Employment rate	Number of employees	N./year
	Green Public Procurement and public-private collaborations	New collaborations between public agencies and enterprises	N./year
		Public tenders incorporating CE and resource efficiency criteria	%
	Circular lifestyle	Number of people using a personal dashboard that display real-time data-flows from smart energy, water and waste bin meters, helping to increase awareness about consumption	N/year
		Percentage of population that shows an increase in circular behaviour	%
		Annual number of visitors (with active engagement) to the reuse hubs	N./year

(continued)

Table 3.1 (continued)

Dimension of indicators	Area of indicators	Type of indicator	Unit of measure
	Health people satisfaction	Percentage of population that describes their own health as good or very good	%/year
		Percentage of population dying from respiratory diseases (due to bad air quality or habits such as smoking)	%/year
	Education level	Population with middle or high education	%/year
	Wealth and poverty	Average household income	€/year
		Population below poverty line	%/year
	Circular initiatives	Number of new circular initiatives	N./year
		Number of local "green" companies	N./year
		Number of new forms of enterprises (SMEs, start-ups, incubators, etc.)	N./year
	Civil CE	Percentage of residents participated in dialogue and/or design related to CE	%/year
	Citizens satisfaction of public services	Level of satisfaction of citizens with the administration services	

The report on material flows marked the starting point for the city's ambition to reduce half of the raw material consumption by 2030 and to be mostly completely circular by 2050.[22] A broad toolset is employed, based around the main themes of raising awareness and anchoring (Fig. 3.2). Much of the efforts are aimed at supporting bottom-up activities originating from entrepreneurs and existing businesses, and cooperating with the port authorities and the Erasmus University. A portal is also provided to spread news regarding the local CE and to provide an overview of all initiatives undertaken by local business and citizens to spread awareness.[23]

3.5 Case Study II: Circular Paris

In line with other cities in the EU, the capital of France, Paris, has officially committed to implementing the CE model since the year 2015. The authorities of the city of Paris with other local authorities of the Greater Paris metropolis organized the General Assembly on the CE. The assembly proceedings led to the publication of a White Paper on the CE. The paper identifies practical proposals in terms of actions and innovative solutions to be applied in the metropolitan territory for pursuing its sustainable development goals.[24]

The CE is part of a larger commitment of the city towards the development of a sustainable, cohesive, responsible, and resilient city. Paris had already adopted some actions (see Footnote 24) in 2014, embedding CE principles such as:

- The zero waste path, aimed to reduce the amount of waste;
- improve recycling and avoiding landfilling;

[22] Rotterdam Circulair, Rotterdam Circularity Programme 2019-2023 https://rotterdamcirculair.nl/wp-content/uploads/2019/05/Rotterdam_Circularity_Programme_2019-2023.pdf.

[23] Rotterdam Circulair, Platform for news and initiatives https://rotterdamcirculair.nl/nieuws/ and https://rotterdamcirculair.nl/initiatieven/.

[24] Paris, Circular Economy Plan 2017-2020 https://cdn.paris.fr/paris/2019/07/24/38de2f4891329bbaf04585ced5fbdf0f.pdf.

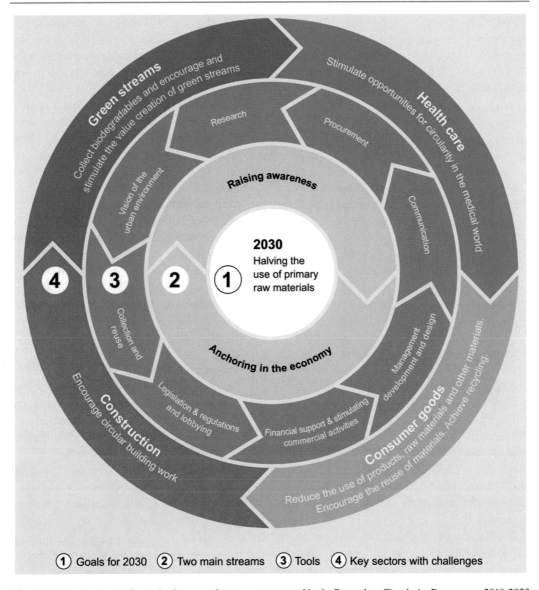

Fig. 3.2 Rotterdam's circular aspirations at a glance, as represented in the Rotterdam Circularity Programme 2019-2023 (see Footnote 22). Reproduced with permission (see Footnote 22); Copyright the Government of the Netherlands

- limit energy recovery to only non-recyclable/non-reusable waste;
- collection of organic waste at the source.

Moreover, within a French national project, Paris had committed to reduce household and similar waste by 10% by the year 2020. In January 2017, Paris adopted a compost plan that included the sorting of waste at the source and the door-to-door collection of household food waste. The city is strongly involved in fighting food wastage by 2025 as promoted in a strategic plan of 2015, containing practical actions identified by consulting local and public stakeholders.

Paris had also started efforts on the public procurement side by the end of 2015, through the initiative of the transnational procurement group that cooperates with several European cities. In February 2016, Paris adopted a responsible public procurement scheme where CE is an important part of the scheme. The scheme's CE principles are indirectly reflected in the commitment of the city to newly defined criteria related to resource efficiency in its public procurement. Additionally, the Paris Climate and Energy Action Plan included many factors common with CE, with targets that are more ambitious than the National and European targets. The city aimed to reduce energy consumption by 25%, and increase consumption of renewable energy by 25% by 2020 (see Footnote 24).

Another area addressed within the envisioned CE perspectives and principles of Paris, has been the re-industrialization of the city by encouraging local production that could fit local needs better and reduce the consumption of materials and energy. In this view, networks for connecting stakeholders have been created (e.g., business and employment centers), logistics projects such as the distribution hubs in the periphery of Paris have been launched, and there is added support to innovation by businesses and research projects. The city also recommends the development of river and rail transport, the growth of innovation hubs, product eco-design and repair, energy-efficient building renovation, the creation of material recovery platforms, and the implementation of short supply chains. Further, the city of Paris organized dedicated events, announcing its goals about the creation of fab labs, maker spaces, and other similar production spaces in the city (see Footnote 24).

The CE Plan of Paris identifies some important CE drivers and recognizes CE as a new approach. It has become part of a more comprehensive territorial project that involves the development of new technologies, new modes of organization between the stakeholders and related sectors, new tools of social innovation as well as new economic models such as stimulating short supply chains. Having a functional and collaborative economy has become a key component. According to this plan, the CE approach needs to be based on the following ten actions (see Footnote 24):

1. A solid territorial plan with specific targets and goals, coherent with other municipal plans, shared and clear for all the stakeholders, and regularly monitored through indicators
2. governance mechanisms involving cooperation between the stakeholders in different sectors and between all stakeholders from public, private, and non-profit sectors
3. cross-cutting organizations capable of adopting more integrated and systemic territory management approaches
4. tools for managing the complexity in addition to the adopted 'Paris urban metabolism' web platform
5. the role of users and their collaboration in the design of service innovations
6. the adoption of an explorative approach to innovations inclusive of their testing, evaluation of operativity, acceptability, and costs
7. an adequate legal framework useful to promote the CE, guaranteeing public-private cooperation and the removal of barriers hampering innovations in the CE transition
8. the existence of incentive-based funding to encourage innovations in the area of CE
9. assessment and organization of CE innovations for their wider application
10. effective and educational communication related to the CE actions.

The Paris CE Plan also presents the following six indicators to measure the impacts of the CE Plan:

1. Tons of waste avoided by means of repair, reuse, recovery, and recycling activities (Tons/year),
2. Tons of waste avoided (Tons/year),
3. Reduction of the size of urban metabolism (input and output flows) (%),
4. Percentage increase in material recovery and organic recovery in Paris (%),
5. number of jobs created,
6. value creation from the development of CE models.

An overview of the CE actions as taken by the municipality of Paris is presented in the below list, which is modified from the Paris CE Plan 2017–2020 (see Footnote 24).

1. **Recovery of green waste**
   i recycling of wood waste from green spaces;
   ii recovery of Christmas trees.
2. **Recovery of materials (from construction and public works)**
   iii recycling of funerary monument dismantling;
   iv recycling of road materials;
   v recovery of materials during large-scale renewal works;
   vi digital interdepartmental exchange platform.
3. **Eco-design of venues and events**
   vii green space eco-design reference;
   viii charter of eco-responsible events.
4. **Sustainable and responsible supplies**
   ix supply of organic and sustainable products in canteens;
   x socially and environmentally responsible public procurement scheme;
   xi development of urban agriculture.
5. **Energy: recovery and reuse of heat and cold**
   xii recovery of heat from wastewater to heat public buildings;
   xiii recovery of heat from data centres;
   xiv cooling of public buildings.
6. **Water management**
   xv development of the non-potable water network and uses;
   xvi rational water management in green spaces.
7. **Mobility and goods transport**
   xvii development of urban logistic spaces;
   xviii shared municipal fleets (cars and bicycles).
8. **Organic waste: separate collection from recovery**
   xix collection of organic waste (municipal restaurant and markets);
   xx launch of separate collection of household food waste;
   xxi support and assistance for collective composting.

9. **Consumer goods: enhancing the extension of the lifecycle**
   xxii recovery of IT and telephony equipment;
   xxiiii reuse of furniture;
   xxiv experimentation with sharing kiosks;
   xxv support for reuse actors (recycling centers, repair cafès, etc.).
10. **Zero waste path: facilitating sorting**
   xxvi deployment of Emmaüs Eco-Systems solidarity collections ;
   xxvii Improving the proximity of bulk waste collection points.
11. **Fight against food wastage**
   xxviii fight against food wastage in municipal canteens;
   xxix recovery of unsold items on food markets;
   xxx support to nonprofit organizations to collect unsold food items.

3.6 Indicators for CE Developed by the EU

The CE as a model of an economic system is aimed at achieving a more sustainable development [5, 46]. This entails sustainable management of natural resources in the society and the economic system for their functioning and uses in human activities [47].

The concept of sustainability in the management of natural resources refers in particular to renewable resources. The management of a resource can be defined as 'sustainable' if its use does not exceed a specific ecological limit coherent with its capacity of reproduction or regeneration in biological cycles, i.e. if not used at a rate that is more than the rate at which nature can produce over the same period [48]. Non-renewable resources, on the contrary, do not have this feature and are defined as exhaustible or finite resources [48]. Fossil fuels can be considered non-renewable, as our global rate of consumption is larger than nature's capacity to produce new fossil fuels (a process that takes ages). Vegetable oils, such as olive oil, are renewable, and using olive oil can be considered sustainable as long as our agricultural systems do not reduce the population of olive trees, nor their ability to produce a constant amount of olives. Sustainability in the management of such resources should reflect longer periods, spanning generations, to assure that both current and future generations are not compromised in meeting their resource needs [48, 49]. One could even consider that economies have to account for projected population growth, to be sustainable if the availability of a resource per capita over multiple generations is factored in. The approach of CE for finite resources suggests their better conservation and replacement with renewable resources.[25] A CE implementation should reduce the demand for natural resources, and reduce environmental impacts. This will allow the decoupling of raw material consumption from economic growth both domestically and abroad, these two factors being typically strongly correlated in a linear economy [22]. This not only requires an increase in recycling and a shift to renewable sources but also absolute reductions in resource extraction and consumption. This implies a downsizing of socio-economic 'metabolism'—the rate of consumption of raw materials [23, 24].

The monitoring framework for the transition to a CE, as set up by the EU, proposed macro-level indicators to shed light on the use of natural resources and the recycling of waste, primarily for the reduction of the demand for natural resources in Europe. Other CE monitoring frameworks have been proposed for the macro level in different countries such as the Netherlands, France, and China [13] The monitoring framework of the EU, despite its limitations, is a useful tool for measuring progress toward the CE, highlighting where the existing policies have been successful and where they are flawed and should be improved [13]. Ten indicators have been identified, they are grouped and represented in Table 3.2.[26]

The indicators are based on the information coming from three main sources: the statistical

office of the EU, named EUROSTAT, the Raw materials scoreboard, and the Resource efficiency scoreboard [13]. Some indicators are still under development, such as those related to Green Public Procurement and Food waste [13]. Other indicators have been consolidated. For example, the recycling rate for the EU of most major waste streams (excluding major mineral waste), has been well documented for each member state. On average, the recycling rate was 56% in 2018, whereas for plastic packaging it was 42%. Additionally, in 2018, about 88% of Construction and demolition waste was recovered and 35% of e-waste was recycled. Another interesting indicator is the circular material use rate (CMU) that measures the share of materials recovered and fed back into the economy as a percentage of overall material use. The CMU was 11,9% for the whole EU (27 countries), in 2019 and it had increased by 3.6% since 2004. At the global level, Haas et al. [24] calculated that the CMU was only 6% in 2005, indicating a very low degree of circularity worldwide. As a result, they suggested the following as key actions:

- the renewable energy transition and the adoption of more preventive measures (such as the reduction of societal stock growth and the application of eco-design for 'circular' products),
- advancing the circularity rate by reducing the consumption of natural resources and to improve (lower) the environmental footprint of the global society [25].

The indicators used by the EU, and most countries actively involved in promoting CE, are quite often focused on recycling at the macro level. The EU's European Environment Agency defines 'circularity' as the average recycling rate of major waste streams.[27] While recycling is indeed a crucial component of any transition to CE, recycling alone is not enough for a society to transition fully to the CE. The EU itself has acknowledged the importance of other waste management methods, such as repair, refurbishment, and remanufacturing, and placed these above recycling in the 'waste

---

[25] European Commission (2020), Circular Economy Indicators, sustainable resource management https://ec.europa.eu/environment/ecoap/indicators/sustainable-resource-management_en.

[26] As reported by EUROSTAT, statistics of the EU https://ec.europa.eu/eurostat/web/circular-economy/indicators/main-tables.

[27] European Environmental Agency (2021), Schematic representation of limits of circularity in the EU-27, 2019 https://www.eea.europa.eu/ds_resolveuid/de949bbcf8e84adbbaaa2176e6483e47.

Table 3.2 Indicators identified in the EU's CE monitoring framework[26]

Thematic areas	Indicators	Sub-indicators (if any)	Unit of measure
Production and consumption	Self-sufficiency of raw materials for production in the EU		%
	Green Public Procurement		
	Waste Generation (as an indicator for consumption activities)	Generation of municipal waste per capita	Kg per capita
		Generation of waste excluding major mineral waste per GDP unit	Kg per thousand euro
		Generation of waste excluding major mineral wastes per domestic materials consumption	%
	Food waste		Million tonne
Waste management	Recycling rates of all waste excluding major mineral waste		%
	Recycling rates of specific waste streams	Recycling rate of municipal solid waste	%
		Recycling rate of packaging waste by type of packaging	%
		Recycling rate of e-waste	%
		Recycling rate of biowaste	%
		Recycling rate of construction and demolition waste	%
Secondary raw materials	Contribution of recycled materials to raw materials demand—end-of life recycling input rates		%
	Circular material use rate		%
	Trade of recyclable raw materials between the EU Member States and with the rest of the world		Tonne
Competitivness and innovation	Private investments, jobs and gross value added related to CE sectors	Gross investments in tangible goods	Million euro
		Number of persons employed	Number
		Value added at factor costs	Million euro
	Patents related to recycling and secondary raw materials as a proxy for innovation		Number

hierarchy in the EU's 2008 Waste Framework Directive.[28] The EU has recently also opened up the debate on degrowth.[29] There is an active discussion to see if GDP is still an adequate indicator of success for a country or economic block that has transitioned to CE, or perhaps other measures of success should be prioritized. The discussion on alternatives is only just starting.

3.7   Concluding Opinion

In this chapter, the authors aimed to provide an overview of the implementation and evaluation of the CE in the macro systems (cities, regions, nations, and beyond). The evaluation methods and

---

[28] European Commission (2008), EU Directive 2008/98/EC on waste (Waste Framework Directive) https://ec.europa.eu/environment/waste/framework/.

[29] European Environmental Agency (2020), Growth without economic growth https://www.eea.europa.eu/ds_resolveuid/beed0c89209641548564b046abcaf43e.

indicators that are currently in use for monitoring the transition to CE were listed. Particular attention was given to cities (through two case studies of cities that have aimed to become circular). A large part of the human population is living in cities in the 21st century, and the urbanization rate is still growing, making the socio-economic metabolism of cities increasingly important. The analysis given includes guidelines for cities to approach the transition to the CE, as well as strategies and indicators already adopted by some cities in the EU and worldwide. In particular, for cities, the multitude of indicators capturing the environmental, social, and economic impacts of the transition reflects the nature of CE. The transition to CE is considered an important socio-technical process towards a wider goal of sustainable development. Environmental indicators are the most numerous and often considered along with economic indicators. Social indicators are receiving comparatively lower attention. This imbalance could be due to the initial stage of development of circularity assessment frameworks and the lack of relevant standards. The authors hope that this study would help increase the understanding of the tools for circularity assessment at the macro level as well as stimulate further wider research and development on social indicators essential for tracking the progress towards a CE, with due consideration given to the well-being of society.

References

1. Corvellec H, Böhm S, Stowell A et al (2020) Introduction to the special issue on the contested realities of the circular economy. Cult Organ 26(2):97–102
2. Lin BCa (2020) Sustainable growth: a circular economy perspective. J Econ Issues 54(2):465–471
3. Cristoni N, Tonelli M (2018) Perceptions of firms participating in a circular economy. Eur J Sust Devel 7(4):105–105
4. Magistris (2020) From programming to implementation: one year after the conference
5. Bauwens T, Hekkert M, Kirchherr J (2020) Circular futures: what will they look like? Ecol Econ 175(106):703
6. Bocken NM, De Pauw I, Bakker C et al (2016) Product design and business model strategies for a circular economy. J Ind Prod Eng 33(5):308–320
7. Fan Y, Fang C (2020) Circular economy development in China-current situation, evaluation and policy implications. Environ Impact Assess Rev 84(106):441
8. Silvestri F, Spigarelli F, Tassinari M (2020) Regional development of circular economy in the European Union: a multidimensional analysis. J Clean Prod 255(120):218
9. Ghisellini P, Ulgiati S (2020) Circular economy transition in Italy. Achievements, perspectives and constraints. J Clean Prod 243:118,360
10. Su B, Heshmati A, Geng Y et al (2013) A review of the circular economy in China: moving from rhetoric to implementation. J Clean Prod 42:215–227
11. Paiho S, Mäki E, Wessberg N et al (2020) Towards circular cities-conceptualizing core aspects. Sustain Urban Areas 59(102):143
12. Liu Z, Adams M, Wen Z et al (2017) Eco-industrial development around the globe: recent progress and continuing challenges. Resour Conserv Recycl 127:A1–A2
13. Moraga G, Huysveld S, Mathieux F et al (2019) Circular economy indicators: what do they measure? Resour Conserv Recycl 146:452–461
14. Corona B, Shen L, Reike D et al (2019) Towards sustainable development through the circular economy-a review and critical assessment on current circularity metrics. Resour Conserv Recycl 151(104):498
15. Fusco Girard L, Nocca F (2019) Moving towards the circular economy/city model: which tools for operationalizing this model? Sustainability 11(22):6253
16. Petit-Boix A, Leipold S (2018) Circular economy in cities: Reviewing how environmental research aligns with local practices. J Clean Prod 195:1270–1281
17. Saidani M, Yannou B, Leroy Y et al (2019) Circular economy in cities: reviewing how environmental research aligns with local practices. J Clean Prod 207:542–559
18. Chen C, Liu G, Meng F et al (2019) Energy consumption and carbon footprint accounting of urban and rural residents in beijing through consumer lifestyle approach. Ecol Ind 98:575–586
19. Geng Y, Sarkis J, Ulgiati S et al (2013) Measuring China's circular economy. Science 339(6127):1526–1527
20. Bosman R, Rotmans J (2016) Transition governance towards a bioeconomy: a comparison of Finland and the Netherlands. Sustainability 8(10):1017
21. Jurgilevich A, Birge T, Kentala-Lehtonen J et al (2016) Transition towards circular economy in the food system. Sustainability 8(1):69
22. Ghisellini P, Cialani C, Ulgiati S (2016) A review on circular economy: the expected transition to a balanced interplay of environmental and economic systems. J Clean Prod 114:11–32
23. Mayer A, Haas W, Wiedenhofer D et al (2019) Measuring progress towards a circular economy: a monitoring framework for economy-wide material loop closing in the EU28. J Ind Ecol 23(1):62–76

24. Haas W, Krausmann F, Wiedenhofer D et al (2015) How circular is the global economy?: An assessment of material flows, waste production, and recycling in the european union and the world in 2005. J Ind Ecol 19(5):765–777

25. Mirabella N, Allacker K, Sala S (2019) Current trends and limitations of life cycle assessment applied to the urban scale: critical analysis and review of selected literature. Int J Life Cycle Ass 24(7):1174–1193

26. Ghisellini, P, Santagata, R, Zucaro, A et al (2019) Circular patterns of waste prevention and recovery. E3S Web Conf 119:00,003

27. Dahlbo H, Bachér J, Lähtinen K et al (2015) Construction and demolition waste management-a holistic evaluation of environmental performance. J Clean Prod 107:333–341

28. Ulgiati S, Ascione M, Bargigli S et al (2011) Material, energy and environmental performance of technological and social systems under a life cycle assessment perspective. Ecol Model 222(1):176–189

29. Huang SL, Hsu WL (2003) Materials flow analysis and emergy evaluation of Taipei's urban construction. Landsc Urban Plan 63(2):61–74

30. Santagata R, Zucaro A, Viglia S et al (2020) Assessing the sustainability of urban eco-systems through emergy-based circular economy indicators. Ecol Ind 109(105):859

31. Caves RW (2005) Encyclopedia of the city. Taylor & Francis, Milton Park

32. Ghisellini P, Casazza M (2016) Evaluating the energy sustainability of urban agriculture towards more resilient urban systems. J Environ Acc Manag 4(2):175–193

33. Casazza M, Liu G, Maglioccola F et al (2020) A retrospective comparison on Europe and China ecological wisdom of pre-industrial urban communities under the lens of sustainability pillars. J Environ Acc Manag 8(4):365–385

34. Torreggiani D, Dall'Ara E, Tassinari P (2012) The urban nature of agriculture: bidirectional trends between city and countryside. Cities 29(6):412–416

35. Zhao X, Huang S, Wang J et al (2020) The impacts of air pollution on human and natural capital in China: a look from a provincial perspective. Ecol Ind 118(106):759

36. Wang Z, Cui C, Peng S (2019) How do urbanization and consumption patterns affect carbon emissions in China? a decomposition analysis. J Clean Prod 211:1201–1208

37. Zhang S, Zhu D (2020) Have countries moved towards sustainable development or not? Definition, criteria, indicators and empirical analysis. J Clean Prod 267(121):929

38. Prendeville S, Cherim E, Bocken N (2018) Circular cities: mapping six cities in transition. Environ Innov Soc Trans 26:171–194

39. Wang N, Lee JCK, Zhang J et al (2018) Evaluation of urban circular economy development: an empirical research of 40 cities in China. J Clean Prod 180:876–887

40. Kisser J, Wirth M (2021) The fabrics of a circular city. In: An introduction to circular economy, pp 55–75

41. Zhu J, Fan C, Shi H et al (2019) Efforts for a circular economy in China: a comprehensive review of policies. J Ind Ecol 23(1):110–118

42. Christis M, Athanassiadis A, Vercalsteren A (2019) Implementation at a city level of circular economy strategies and climate change mitigation - the case of Brussels. J Clean Prod 218:511–520

43. Huovila A, Bosch P, Airaksinen M (2019) Comparative analysis of standardized indicators for smart sustainable cities: what indicators and standards to use and when? Cities 89:141–153

44. Cavaleiro de Ferreira A, Fuso-Nerini F (2019) A framework for implementing and tracking circular economy in cities: the case of porto. Sustainability 11(6):1813

45. Endreny TA (2018) Strategically growing the urban forest will improve our world. Nat Commun 9(1):1–3

46. Kirchherr J, Reike D, Hekkert M (2017) Conceptualizing the circular economy: an analysis of 114 definitions. Resour Conserv Recycl 127:221–232

47. Koltun P (2010) Materials and sustainable development. Progress Nat Sci: Mater Int 20:16–29

48. Stahel WR (2016) The circular economy. Nature News 531(7595):435

49. Lanza A (2002) Sustainable development. Il Mulino

Circularity at Meso Level: A Sector Perspective

Patrizia Ghisellini, Rashmi Anoop Patil, Sven Kevin van Langen and Seeram Ramakrishna

Abstract

This chapter provides an overview of the application of CE at the meso level. The authors focus on EIPs, as the most representative cases at this level. The EIPs are unique examples showing how companies can cooperate in sharing resources such as water, energy, material, by-products, and services. Case studies of EIPs (IZ NÖ-Süd in Austria and Ulsan Mipo and Onsan Industrial Park) are presented, with their environmental, economic, and social performances tracked through existing evaluation frameworks. To provide a deeper perspective on the topic, the origins, evolution, and current performances of Kalundborg Symbiosis EIP in Denmark, the well-known longstanding case of EIP are briefly summarized. Moreover, the current monitoring framework developed by the Chinese government complements the analysis. So far, China has developed the largest EIP program worldwide. Overall, the EIPs case studies show that, besides the economic benefits, EIPs provide environmental and social benefits depending on their implementation and management. In such a way, the cases show how EIPs can contribute to diversifying the industrial context in a more sustainable way and more in harmony with the natural environment and the surrounding social community.

P. Ghisellini(✉)
Department of Engineering, Parthenope University of Naples, 80143 Naples, Italy
e-mail: patrizia.ghisellini@gmail.com

R. A. Patil · S. Ramakrishna(✉)
The Circular Economy Task Force, National University of Singapore, Singapore 117575, Singapore
e-mail: seeram@nus.edu.sg

S. K. van Langen
UNESCO Chair in Environment, Resources and Sustainable Development (International Ph.D. Programme), Department of Science and Technology, Parthenope University of Naples, 80143 Naples, Italy

Olympia Electronics, Thessaloniki, Greece

S. Ramakrishna
Department of Mechanical Engineering, National University of Singapore, Singapore 117575, Singapore

Centre for Nanotechnology and Sustainability (NUSCNS), 2 Engineering Drive 3, Singapore 117576, Singapore

Keywords

Meso-level circularity assessment · Environmental sustainability · Sector circularity · Circular industrial parks

© The Author(s) 2023
R. A. Patil and S. Ramakrishna (eds.), *Circularity Assessment: Macro to Nano*,
https://doi.org/10.1007/978-981-19-9700-6_4

4.1   Introduction

One of the core aspects or "building blocks" of the model of CE, that revolutionized the previous linear economy model, is the design of circular production and consumption models that aims to reducing, slowing and closing the flows of resources [1]. As such "reducing" reflects in general in improving the efficiency in the use of resources needed for products, processes, or systems whereas the "slowing" also reduces the speed in the use of resources by designing long-life products (e.g., applying the concept of design for durability or reliability) as well as extending their service life (through strategies such as reuse, maintenance, repair, and technical upgrading). The last strategy "closing the loops" recycles materials or other kinds of resources (e.g., by-products) by closing the loops in both post-production and post-consumption stages avoiding landfilling[1] [1]. At the company or industry level, the application of CE translates into the adoption of cleaner production processes creating the opportunity of exploring internal and external recycling processes with other companies in the supply chain [2–4]. Through internal or external recycling strategies, the industrial activities operate more closely to the functioning of natural ecosystems where resources are never considered as a "waste" [5]. Frosch and Gallopoulous [6] in their seminal work emphasized the need for industrial activities to be more integrated by cooperating in exchanges of by-products and resources. They quote that:

The traditional model of industrial activity... should be transformed into a more integrated model: an industrial ecosystem. In such a system the consumption of energy and materials is optimized, waste generation is minimized, and the effluents of one process...serve as the raw material for another [6].

These concepts, on the production side, translate into cooperative networks of resource exchanges (materials, water, energy, and by-products) between independent companies of the same sector or other sectors [3]. The primary essence of establishing such resource exchange for "industrial symbiosis" [7] is gaining an economic advantage, of course, there are environmental benefits too [7]. In this regard, the Kalundborg eco-industrial park in Denmark is a reference case all over the world for a smart community of companies applying circularity.[2] Other EIPs have been identified all over the world, analyzed, and documented well in literature [7–10].

The meso level of CE implementation also comprises programs of residential complexes of households aimed to reduce and optimize the use of energy, water, and solid waste [11, 12]. Moreover, resource exchanges can be realized in cities between the civil society (e.g., households) and private sector (e.g., industries) with the support of local government. Applications of these practices can be found in Rotterdam (The Netherlands) and Japanese eco-cities [9, 13]. The Rotterdam Energy Approach and Planning (REAP) started in 2009 by the Rotterdam local authority's aims to realize several initiatives of urban symbiosis at different levels: city, district, neighborhood, and building level. At the city level, it involves the reuse of waste energy flows coming from the harbor industries towards the district heating grid. At the district/neighborhood scale, the waste heat of offices and shops can be directed to homes where the energy can be exchanged between swimming pools (requiring heat) and ice-skating rings (requiring cooling and using renewable energy).[3]

In this chapter, we begin by providing definitions and features of EIPs and further present the assessment frameworks and indicators useful for industrial parks in transitioning to the model of EIP as well as for evaluating the performance of EIPs with illustrations and case studies (IZ NÖ-Süd in Austria and the Ulsan Mipo and Onsan Industrial Park in South Korea). Then, the

---

[1] Ellen Mac Arthur Foundation https://www.ellenmacarthur-foundation.org/case-studies/business/building-blocks/reverse-cycle.

[2] Kalundborg Symbiosis: six decades of a circular approach to production https://circulareconomy.europa.eu/platform/fr/node/938.

[3] Urban Symbiosis: Recommendations for Cities to Re-use Resources https://ec.europa.eu/environment/ecoap/news/urban-symbiosis-recommendations-cities-re-use-resources_en.

National Eco-industrial Park Standard in China inclusive of the indicators adopted for evaluation is provided as a further case study. Also, the Kalundborg EIP is discussed as another case that includes a description of its environmental and socio-economic performances. We conclude with a summary of this chapter on Eco-Industrial Parks.

4.2 Understanding Eco-Industrial Parks (EIP)

One of the first definitions of Eco-industrial park has been provided by Cotè and Hall [14, 15] as follows: *an industrial system which conserves natural and economic resources; reduces production, material, energy, insurance and treatments costs and liabilities; improves operating efficiency, quality, worker health and public image; and provides opportunities for income generation from use and sale of wasted materials.* This definition has been further expanded by Lowe [16] to emphasize that the aim of realizing cooperative strategies for the company's part of an EIP is the realization of particular synergies. The EIPs show that businesses are not always playing a zero-sum game, by working together to achieve more benefits (economic, environmental and social) as compared to the case in which each company worked alone. In the context of EIP as a whole, some authors compare the behavior of companies interacting with each other to a chemical reaction, where the combination of the reagents gives a product and some residues in the form of waste and emissions. However, for this to happen, the reaction should have some activation energy, be profitable, and release more energy than required for activating the reaction [17]. Similarly, the rationale underlying the realization of an EIP is to find companies (reagents) that are highly compatible (reactive with one another) in terms of input and output, and the procedures to realize cooperative strategies that are profitable for the participating companies [17]. The implementation of EIPs can be the result of planned projects defined at the national level as in the case of China, that has so far developed the largest

EIP program [18, 19]. However, there are also EIPs that have been created based on pre-existing spontaneous initiatives of symbiotic relationships among the participating companies (as in the case of Kalundborg in Denmark) and mixed experiences of EIPs (Burnside, Kawasaki, Central Gulf Coast, Kwinana) where the top-down planning of the EIP project is associated with the unplanned nature of the symbiotic relationships occurring in the EIP [18, 20].

Cotè and Cohen-Rosenthal defined the key elements characterizing an EIP to distinguish the EIP from a conventional industrial park [15], illustrated in Fig. 4.1. These include (adapted from Cotè and Cohen-Rosenthal [15]):

1. The definition of the stakeholders' community of the EIP and the consideration of their involvement in the design of the park;
2. The reduction of environmental impacts and the ecological footprint by replacing toxic materials, absorption of CO_2, material exchanges, and integrated treatment of wastes;
3. Maximization of the energy efficiency by utilizing the design of facility and construction, co-generation and cascading;
4. Conservation of materials through facility design and construction, reuse, recovery and recycling;
5. Link or network of companies with suppliers and customers in the wider community in which the eco-industrial park is located;
6. The continuous improvement of the environmental performances by the individual businesses and the community as a whole;
7. A flexible regulatory system that encourages the companies to meet the performance goals;
8. The use of economic instruments that encourage waste and pollution prevention;
9. The adoption of an information management system (monitoring system) that supports the flow of energy and materials within a more or less closed-loop;
10. The creation of a mechanism for the training and education of managers and workers about new strategies, tools, and technologies to improve the system;

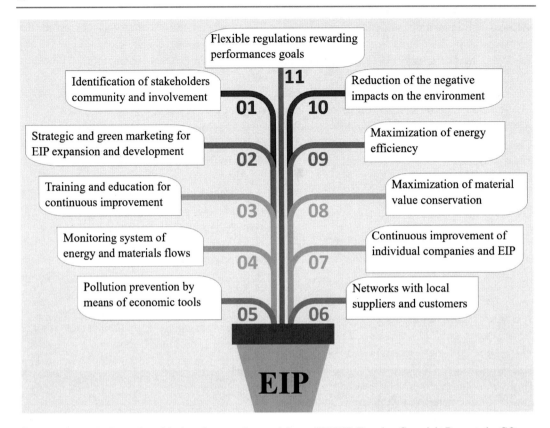

Fig. 4.1 Schematic illustration of the key elements characterizing an EIP [15]. Template Copyright PresentationGO.com

11. The orientation of the marketing (including green marketing) of the EIP, to favor the inclusion of new companies that fill niches and complement other businesses [15].

4.2.1 The Need for Evaluating the Performances of EIPs

The development of EIPs tries to address the need for mitigating the environmental impacts of industrial areas [21]. The classification of EIPs stems from the presence of a community of companies that cooperate with each other in the sharing/exchange of resources (materials, water, energy, by-products) and/or infrastructure. The EIPs are then helpful in achieving greater resource efficiency through the realization of 'economies of systems integration' where a

central role is played by the adoption of IS. The latter involves the activation of complex interplay of resource exchanges (materials, water, energy, and by-products) within the participating companies to achieve socio-economic and environmental benefits [16, 17].[4] *The essence of industrial symbiosis is taking full advantage of by-product utilization, while reducing residual products or treating them effectively. The term is usually applied to a network of independent companies that exchange by-products and possibly share other common resources* [23]. Economic benefits arise for example from avoiding the

---

[4] Please refer Chertow [22] for further learning on industrial symbiosis and related concepts such as 'kernel' and 'precursor'. The words 'kernel' and 'precursor' describe bilateral or multilateral exchanges having only the potential to expand into industrial symbiosis since they do not yet meet the minimum criterion for being defined as industrial symbiosis introduced by Chertow and her colleagues as '3-2 heuristic'.

waste disposal costs and the purchasing of raw materials while the environmental benefits come from the reduction of waste generation and the exploitation of new resource inputs coming from participating companies [18] substituting them with commercial products or raw materials [17].

The EIPs can be evaluated from different perspectives. For example, in terms of performance requirements for existing industrial parks to shift towards the EIP model (such as the UNIDO EIP toolkit and presented in Sect. 4.3). Integral models are also proposed for increasing the knowledge of the metabolism of EIPs and improving their management under the CE perspective [24]. The EIPs can also be evaluated using well-known assessment methods (for example, MFA, LCA, EMA) and related indicators.

For example, [25] applies the MFA to quantify the generation, management (reuse or recycling both on-site and off-site), and disposal of industrial waste in two industrial estates in the Nanjangud Industrial Area (India). The authors evidence that MFA is a key method in industrial ecology research for evaluating the metabolism of a defined system (be it an individual facility or an industrial area) in its input and output including waste generation, and their recovery inside companies, within the companies and outside the EIP. Such a method can also be useful in suggesting potential alternative uses of waste materials and evaluate the improvements in resource efficiency [26]. Other authors used LCA to identify existing and potential industrial symbiosis opportunities and their environmental impacts and benefits. Dont et al. [27] used a hybrid LCA to assess the carbon footprint of the Shenyang Economic and Technological Development Zone, an EIP in China for identifying the sources (on-site and off-site the EIP) and industries contributing to the highest environmental impacts. They found that almost 45% of the latter come from on-site activities of the EIP whereas 55% of them from off-site the EIP. The Emergy Accounting has also been applied to analyze the performances of SETDZ EIP in China in terms of environmental and economic benefits coming from the existing IS. The authors also identified opportunities of increasing the benefits by the expansion of exchanges towards those related to the reuse of treated wastewater from local wastewater treatment, the reuse of sludge from wastewater treatment as a fertilizer, the reduction of the use of coal as an energy source and its substitution with renewable energies such as wind energy [28]. Finally, Chertow et al. [29] studied the number of by-products that could potentially be reused in industrial symbiosis in the Mysuru industrial district in India to understand the potential public benefits (e.g. saved landfill capacity, lower need for public wastewater treatment) deriving from the potential activation of the IS and also, which industries could be facilitated in the activation of the IS by policymakers. They also assessed using a Life Cycle Impact Assessment, the potential impacts of the IS, finding relevant environmental benefits in terms of reduction of CO_2 and PM_{10} in the IS scenario compared to that without IS.

4.3 Evaluating Eco-Industrial Parks: The Assessment Framework by UNIDO

The UNIDO EIP toolkit[5] has been developed for guiding industrial parks towards the mainstreaming of EIPs and assuring their compliance to sustainability. In that, it has been intended as a useful tool for supporting the implementation and decision-making process of existing and new industrial parks. The EIP Toolboxprovides a wide range of tools useful for the selection of industrial parks for new EIP projects, stakeholder mapping, policy support, assessing industrial parks, identification of symbiotic industries, monitoring impacts from company-level production activities, and park-level EIP opportunities [30].

The various categories and the major aspects covered under the UNIDO toolkit are given in Table 4.1 that lists the performance requirements for EIPs within four key categories: (i) park man-

---

[5] The toolkit has also been included in the collaboration of UNIDO, The World Bank, and the Deutsche Gesellschaft fur Internationale Zusammenarbeit GmbH (GIZ) (German Development Cooperation) for supporting the development of EIPs worldwide. https://openknowledge. worldbank.org/handle/10986/29110.

Table 4.1 Park Management, Environmental, Economic, and Social performances requirements for EIPs (see Footnote 5). *Source* The UNIDO, the World Bank Group, and Deutsche Gesellschaft fur Internationale Zusammenarbeit (German Development Cooperation) (GIZ) GmbH, 2017

Category	Sub-category	Description of indicators	Unit (Target value)
Park management services	Park management empowerment	Proportion of the firm that has signed a residency, contract/park, charter/code of conduct and further legally binding arrangements that empower the park management entity to perform its responsibilities and tasks, and charge fees	% of firms
	Park management entity, property and common infrastructure	The resident firms indicate satisfaction with regard to the provision of services and common infrastructure by the park management's entity (or alternative agency where applicable)	% of firms
Monitoring and risk management	EIP performance and critical risk management	Park management entity regularly monitors and prepares consolidated reports regarding the achievement of target values to encompass the environmental, social and economic performances and critical risk management at the level of the park	Frequency of reports
Management and Monitoring	Environmental/Energy Management Systems (EMS and EnMS, respectively)	Proportion of resident firms, with more than 250 employees, having an environmental/energy management system in place that is in line with internationally certified standard	Percentage of firms (40%)
Energy	Energy consumption	Proportion of combined park facilities and firm-level energy consumption, for which metering and monitoring systems are in place	Percentage of combined park and firm level energy consumption (90%)
	Renewable and clean energy	Total renewable energy use in the industrial park is equal to or greater than the annual national average energy mix	Percentage of renewable energy use in park relative to national average % ($\geq$)
	Energy efficiency	Park management entity sets and works toward ambitious (beyond industry norms) maximum carbon intensity targets (maximum kilograms of carbon dioxide equivalents (kg CO_2 eq./kWh) for the park and its residents. Targets should be established for the short, medium and long term	kg CO_2e/kWh (in line with local norms and industry sector benchmarks)
		Park management entity sets and works toward ambitious maximum energy intensity targets per production unit (kWh/$ turnover) for the park and its residents. Targets should be established for the short, medium and long term	kWh/$ turnover norms and industry sector benchmarks)

(continued)

Table 4.1 (continued)

Category	Sub-category	Description of indicators	Unit (Target value)
Water	Water consumption	Total water demand from firms in industrial park which do not have significant negative impacts on local water sources or local communities	Percentage of water demand (100%)
	Water treatment	Proportion of industrial wastewater generated by industrial park and resident firms, which is treated to appropriate environmental standards	Percentage of waste water treated/total waste water (95%)
	Water efficiency, reuse and recycling	Proportion of total industrial wastewater from firms in the park are reused responsibly within or outside the industrial park	Percentage of water reused/total water consumed (50%)
Waste and material use	Waste/by product reuse and recycling	Proportion of solid waste generated by firms, which is reused by other firms, neighboring communities, or municipalities	Percentage of solid waste reused/total waste (20%)
	Dangerous and toxic materials	Proportion of firms in park, which appropriately handle, store, transport and dispose of toxic and hazardous materials	Percentage of firms (100%)
	Waste disposal	Maximum proportion of wastes generated by firms in the industrial park which go to landfills	Percentage of waste to landfill (<50%)
Climate change and the natural environment	Flora and fauna	Minimum proportion of open space in the park used for native flora and fauna	Percentage of open space (5%)
	Air, Greenhouse Gas (GHG) emissions and pollution prevention	Proportion of firms in park which have pollution prevention and emission reduction strategies to reduce the intensity and mass flow of pollution/emission release beyond natural regulations	Percentage of firms (50%)
		Proportion of largest polluters in industrial park which have a risk management framework in place that: (a) identifies the aspects which have an impact on the environment and (b) assign a level of significance to each environmental aspect	Percentage of largest emitters (30%)
Social management system	OH&S management system	Percentage of all firms in the industrial park with more than 250 employees that have a well-functioning OH&S management system in place	Percentage of firms (75%)
	Grievance management	Percentage of grievances received by the park management entity which are addressed within 90 days	Percentage of grievances (100%)
		Percentage of grievances received by the park management entity, which were brought to conclusion	Percentage of grievances (60%)
		Percentage of all firms in the industrial park with more than 250 employees that have a code of conduct system in place to deal with grievance	Percentage of firms (75%)

(continued)

Table 4.1 (continued)

Category	Sub-category	Description of indicators	Unit (Target value)
	Waste disposal	Maximum proportion of wastes generated by firms in the industrial park which go to landfills	Percentage of firms (75%)
Social infrastructure	Primary social infrastructure	Percentage of the surveyed employees reporting satisfaction with social infrastructure	Percentage of surveyed employees (80%)
	Industrial park security	Percentage of reported security and safety issues that are adequately addressed within 30 days	Percentage of reported security and safety issues (100%)
	Capacity building	Percentage of all firms in the industrial park with more than 250 employees having a program for skills/vocational training and development	Percentage of firms (75%)
		Percentage of female workforce who benefit from available supporting infrastructure/programs for skills development	Percentage of female workforce ($\geq$20%)
Local community outreach	Community dialogue	Over 80% of the surveyed community members are satisfied with the community dialogue	Percentage of surveyed community members (80%)
	Community outreach	Number of outreach activities implemented by the park management entity annually that are regarded as positive by over 80% of the surveyed community members	Number of outreach activities per year
Employment generation	Local employment generation	Percentage of total workers employed in industrial park who live within daily commuting distance	Percentage of employees (60%)
	Type of employment	Percentage of total firm workers in industrial park employed through direct employment (that is, not employed on a fee-for-output basis or provided through a labor supply firm) and permanent contracts	Percentage of employees (25%)
Local business and SME promotion	Local value added	Percentage of resident firms using local suppliers or service providers for at least 80% of their total procurement value	Percentage of firms (25%)
		Percentage of total procurement value of park management entity supplied by local firms or service providers	Percentage of total procurement value of park management entity (90%)
Economic value creation	Investment-ready park for firms	The ratio of rented or used space by resident firms compared to the total amount of available space marked for resident firms within IPs	Average percent occupancy rate over 5 years (50%)

agement performance, (ii) environmental performance, (iii) social performance, and (iv) economic performance. The framework provides the basis for defining and setting prerequisites and perfor-mance requirements for EIPs, based on 51 criteria (benchmarks) [30]. However, it is important to underline that the development of the EIP toolkit aims not only to define minimum performance requirements to be met by industrial parks in transitioning to the model of EIP but to hopefully stimulate the EIP for the continuous improvement of their performances.

The UNIDO, The World Bank, and GIZ under their project and collaborative efforts in supporting the EIPs development worldwide have published a working paper, in which they describe in detail, the just above-mentioned toolkit and apply it to the evaluation of the performances of existing EIPs. This is to show how the adoption of the concept of EIPs in practice provides the case for more sustainable and inclusive development of industrial parks. Here, we report the main results of the analysis (mainly qualitative as only a few quantitative data are provided) from such a working paper on two existing EIPs[6]: the industrial IZ NÖ-Süd in Austria and the Ulsan Mipo and Onsan Industrial Park in South Korea. A brief explanation of the framework implementation for the IZ NÖ-Süd and the Ulsan Mipo and Onsan Industrial Park has been provided below (see Footnotes 5 and 6).

4.3.1 Industrial Zone NÖ-Süd Eco-Industrial Park, Austria

The IZ NÖ-Süd EIP in Austria is more than half a century old and comprises 370 companies. Most of them are SMEs as well as international companies that mainly rent the facilities for office, storage, and production space purposes. The sectors covered by the companies include food and beverage, aluminum and steel converting,[7] production of energy and technical components, environmental services and technologies, and logistics. The EIP is managed by a private business holding

company named 'Ecoplus' which is experienced in managing EIPs. Such a private business aims to ensure the achievement of an added value for the region, the creation of local jobs, and a more sustainable regional development. The compliance to the above requirements translates into the following EIP performances and impacts:

1. **Park Management:** The holding company Ecoplus acts as a hub that links institutions, public authorities, and partners. It supports companies from the creation of their business idea to its financing. Ecoplus also supports companies in the EIP in facilitating relations with local authorities, for example, in obtaining permits for the company's activities.

2. **Economic performance:** The Ecoplus business park IZ NÖ-Süd employs around 11,000 people and relies on long-term collaborations with local vocational schools in the neighboring 4 municipalities (namely, Biedermannsdorf, Guntramsdorf, Laxenburg, and Wiener Neudorf). The collaboration is useful for the recruitment and retention of skilled work force. Ecoplus provides other economic core services, including the creation of business networks, organization of conferences and event facilities, coordination of joint media initiatives for companies and the EIP. Additionally, Ecoplus collaborates with universities to better address the issues of industrial development and its environmental and social sustainability.

3. **Environmental performance:** Ecoplus operates and provides central infrastructure services for the EIP such as a central wastewater treatment plant (totally renovated in 2015–2017), 17 km of access roads, and bus routes, rail connections, and a freight station with the Austrian railroad. Additionally, Ecoplus maintains 100,000 square meters (m^2) of green spaces, shrubs, and trees within the industrial parks. This positive landscaping provides space for recreational activities.

4. **Social performance:** There is an extensive social infrastructure provision in and around the EIP, enabling the growth of a small city. This offers easy access to postal offices and

---

[6] An International Framework For Eco-Industrial Parks https://openknowledge.worldbank.org/bitstream/handle/10986/291 10/122179-WP-PUBLIC-AnInternationalFrameworkforEcoIndustrial Parks.pdf?sequence=1\&isAllowed=y.

[7] Converting is a type of metallurgical smelting that includes several processes; the most commercially important form is the treatment of molten metal sulfides to produce crude metal and slag, as in the case of copper and nickel converting. https://en.wikipedia.org/wiki/Converting_(metallurgy).

custom services for shipments, restaurants, shopping malls, child care facility, security system, well-designed navigation system to guide visitors through the EIP. Furthermore, space for recreational activities dedicated to employees and local communities are provided including tennis courts and golf courses.

4.3.2 Ulsan Mipo and Onsan Eco-Industrial Park, South Korea

The Ulsan Mipo and Onsan EIP in South Korea originated from the transformation of the Mipo-Onsan conventional national industrial complexes into a more sustainable EIP operating according to the national EIP development master plan. The EIP is developed over an area of 6,540 hectares and involves currently 1,000 companies. The companies operate in various industrial sectors such as automobile manufacturing, shipbuilding, oil refining, machinery manufacturing, non-ferrous metals, fertilizer, and chemical industries. Overall, the EIP employs more than 100,000 people and has supported inter-business synergies on the basis of the outputs and requirements of each of the businesses in the EIP network [31]. A typical example of such a synergy [31] appears in Fig. 4.2.

The Ulsan EIP Centre is in charge of the management of the EIP, the reception of the project proposals as well as monitoring the achievements of the EIP in terms of economic, environmental, and social benefits. Some results of such monitoring activity are the following:

1. **Economic performance:** The economic benefits arise in the form of cost savings (resource procurement, operations, and environmental/waste management by replacing virgin materials with by-products) and revenues (generated by selling by-products) which are annually reported. Relevant government investments support research projects and development, including center operations. The government funds contribute to funding new research projects that generate new added value from selling by-products and waste for recycling purposes. Further

benefits come from energy and material savings.[8]

2. **Environmental performance:** The monitoring of the environmental benefits resulted in the reduction of energy consumption, as well as a reduction in the generation of waste or by-products, wastewater, and CO_2 emissions. During 2005–2016, the Ulsan EIP program saved energy equivalent of 279,761 tons of oil. This contributed to the reduction of 665,712 tons of CO_2 emissions and 4052 tons of toxic gases, such as SO_x and NO_x. In addition, the reuse of 40,044 tons of by-products and waste contributed to improving the negative image of the industrial complexes related to the emission of pollutants and the social relations with the neighboring communities.

3. **Social performance:** Compared to the previous EIP, the social performances are not well documented as the data only evidence the economic investments for the construction of networking facilities for industrial symbiosis (totaling US\$ 245.8 million in the year 2016) and the creation of 195 new jobs.

4.4 Case Study I: The National EIPs Evaluation Standard System of China

Since the beginning of this millennium, China has developed the largest global EIPs program. Currently, the Chinese Ministry of Environmental Protection has approved more than a hundred EIP projects [8, 32, 33]. Three main types of EIPs included in the program are sector-integrated EIPs, sector-specific EIPs, and venous industry EIPs.

The Chinese EIPs program relies on the development of a National EIPs evaluation standard system introduced in the year 2006 (updated in 2015). Also, China happens to be the first and only country in the world to have a National EIPs eval-

---

[8] An International Framework For Eco-Industrial Parks https://openknowledge.worldbank.org/bitstream/handle/10986/29110 /122179-WP-PUBLIC-AnInternationalFrameworkforEcoIndustrial Parks.pdf?.

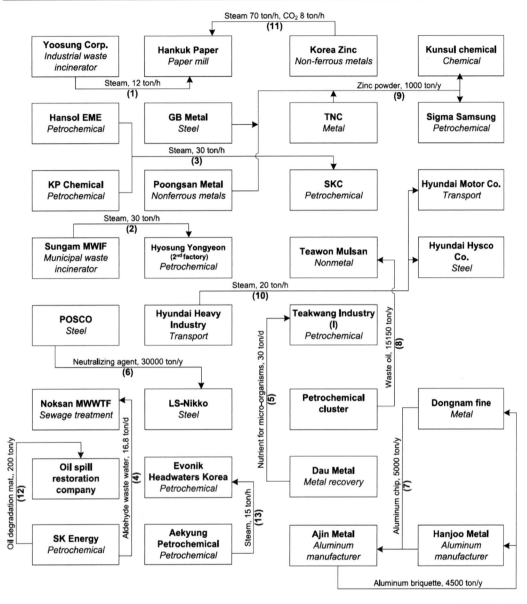

Fig. 4.2 An overview of symbiosis developed in the Ulsan EIP. The network of arrows connecting varies industries in the park, are channels for delivering by-products/waste from one company which is useful to the neighboring one. Reproduced from [31]. Copyright, 2012 Elsevier

uation standard system. As evidenced in Table 4.2 (adapted from [33]), the standard includes five categories of indicators related to economic development, industrial symbiosis, resource conservation, environmental protection, and information disclosure. Each of the indicators' categories is associated with specific targets whose requirements should be satisfied by the EIPs.

Compared to the previous versions of the standard, the latest one applies to all the three types of EIPs with more stringent indicator thresholds (as mentioned above) as there is little use of specific indicators for each one of the three types of EIPs. The industrial symbiosis category of indicators was included for driving the adoption of further industrial symbiosis with others par-

Table 4.2 Performance Indicators in the National Chinese EIP standard system. Adapted from [33]

Category	No.	Indicators	Units
Economic development	1	The proportion of high-tech enterprises output value of gross industrial output value	%
	2	Industrial added value per capita	10^4/person
	3	The average three-year growth rate of industrial added value	%
	4	The proportion of remanufacturing industry added value of the gross industrial added value	%
Industrial symbiosis	5	The added eco-industrial chain numbers after enforcing EIP demonstration program	Units
	6	Comprehensive utilization rate of industrial solid waste	%
	7	Usage rate of renewable resources	%
Resource conservation	8	Industrial added value per unit industrial land area	100 million/km^2
	9	The average three-year annual growth rate of industrial added value per unit industrial land area	%
	10	Elastic coefficient of comprehensive energy consumption	–
	11	Energy consumption per unit of industrial added value	Metric ton of standard coal/10^4 RMB
	12	Application ratio of Renewable energy	%
	13	Elastic coefficient of fresh water consumption	–
	14	Fresh water consumption per unit industrial added value	m^3/10^4 RMB
	15	Recycling rate of industrial water	%
	16	Reuse rate of reclaimed water	%
Environmental protection	17	Rate of reaching the discharging standard for key pollution sources	%
	18	The conditions of national and local key pollutant emissions	–
	19	Frequency of severe environmental accidents	–
	20	Completion degree of environmental management strategies	%
	21	Implementation rate of key enterprises' clean production audit	%
	22	Centralized sewage treatment facilities	–
	23	The completion rate of environmental risk prevention and control system	%
	24	Utilization rate of industrial solid waste (including hazardous wastes)	%
	25	Elastic coefficient of main pollutant emissions	–
	26	The annual reduction rate of carbon dioxide emissions per unit industrial added value	%
	27	Wastewater emission per unit industrial added value	T/10^4 RMB
	28	Solid waste discharge per unit industrial added value	T/10^4 RMB
	29	Green cover percentage	%

(continued)

Table 4.2 (continued)

Category	Sub-category	Description of indicators	Unit (Target value)
Information disclosure	30	Environmental information disclosure rate of key enterprises	%
	31	The completion degree of the ecological industry information platform	%
	32	Number of public education campaigns	No./year

ticipants. Moreover, another key aspect is the inclusion of the indicator "usage rate of renewable resources" that supports the regeneration and reuse of renewable resources within the existing network of industrial symbiosis. Additionally, the current standard includes environmental risk control indicators for better management of hazardous materials and prevention of environmental accidents. The standard also includes environmental indicators such as "elasticity coefficient of main pollutant emissions" for evaluating the opportunity of contributing to the decoupling between resource consumption and emissions of pollutants and economic growth [33].

Proposals for future improvements consider the expansion of industrial symbiosis indicators for understanding better, the practical implications (for example, economic benefits of the industrial symbiosis), and the sharing of other resources (such as waste heat recovery). The proposal also includes ideas for taking into account the impacts of the EIP on its surroundings and its contribution to promoting a more sustainable local development. In this view, the adoption of exchange of resources with the local community surrounding the EIPs is relevant and the inclusion of social indicators for evaluating and monitoring the social impacts of EIPs can be expected. The role of stakeholders (for example, employees, local government, and communities) is key for the success of participant companies and the whole EIP towards a smarter and longstanding perspective [33].

4.5   Case Study II: Kalundborg Symbiosis in Denmark

The Kalundborg Industrial Symbiosis,[9] Denmark, illustrated in Fig. 4.3 is regarded as the first bottom-up example of industrial symbiosis, that began in 1961 from the spontaneous initiative of a few companies whose initial resource exchange comprised the area of water supply. The number of companies involved (inclusive of those belonging to the heavy industries sector) grew progressively over time, subsequently increasing the exchange of resources, rendering in this way the CE challenge a more accessible reality [5]. This setup is in complete contrast with the traditional vision of industrial systems of being independent and competitive [15]. The industrial symbiosis at Kalundborg is an example of a local public-private partnership where the participant companies provide, share, and reuse energy, water, and materials to create new and mutual economic benefits. The economic benefits have been a key driver, guiding over such a long time, the adoption of new projects and their advancement [18] even if other factors such as those related to resource scarcity (e.g. low water availability) have been relevant in each of the resource exchange projects [34]. The business models of circular start-ups seem the most suitable in Kalundborg due to the existence of industrial symbiosis and the continuous innovation in this context. Besides cooperation between participants, crucial factors such as trust, confidentiality, and openness make Kalundborg a successful and longstanding case study and a

---

[9] Explore the Kalundborg Symbiosis; Last accessed: 31/07/2021 http://www.symbiosis.dk/en/.

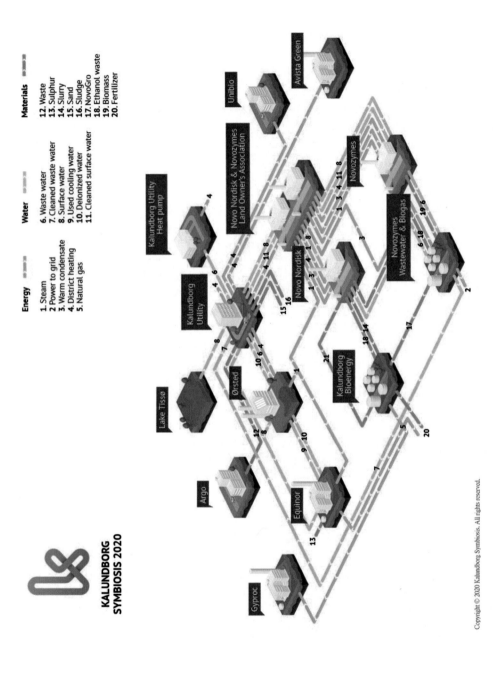

KALUNDBORG SYMBIOSIS 2020

Energy

1. Steam
2. Power to grid
3. Warm condensate
4. District heating
5. Natural gas

Water

6. Waste water
7. Cleaned waste water
8. Surface water
9. Used cooling water
10. Deionized water
11. Cleaned surface water

Materials

12. Waste
13. Sulphur
14. Slurry
15. Sand
16. Sludge
17. NovoGro
18. Ethanol waste
19. Biomass
20. Fertilizer

Fig. 4.3 Kalunborg eco-industrial park with the participant companies and the tracking of energy, water and materials exchanges among them (Courtesy of the Administrative Project and Communications Manager of Kalundborg Symbiosis) (see Footnote 8). Copyright 2020, European Union

real icon of the industrial ecosystem at the global level.

Moreover, this EIP (formalized in 2011 as a private association named 'Kalundborg Symbiosis') relies on a structure inclusive of a board of directors comprising a member from each of the participant companies and regular meeting schedule needed to take the relevant decisions such as the implementation of new symbiosis projects.[10] In terms of structure and functioning, the Kalundborg Symbiosis consists of 25 different resource flow channels for water, energy, and material flows, linking six industrial sectors and three public sector organizations.

The combined annual benefits for the partners in Kalundborg have been assessed through LCA taking into account the data flows in 2015. In monetary terms, it amounted to 24 million euros along with socio-economic benefits (for example, those derived from avoiding costs in waste management) amounting to 14 million euros (see Footnote 9). The combined annual environmental benefits for the partners of the Kalundborg Symbiosis are as follows:

- Reduction of 635,000 tons of CO_2
- Reduction of 3, 6 million m^3 of water
- Reduction of 100 GWh of energy
- Reduction of 87,000 tons of materials.[11]

Many studies have analyzed the origins, functioning, and performances of Kalundborg, along with providing qualitative and quantitative evidence of the exchanges occurring in such EIPs [34] highlighting the economic and environmental benefits. For example, there are exchanges of high-temperature steam from Ørsted's combined heat and power plant (located at Enghave Brygge, Sydhavnen in Copenhagen, Denmark), to many of the other partners in the symbiosis (see Footnote 8), saving energy and benefitting the companies at the same time. A further study [35] analyzed the evolution of the resilience capacity of Kalundborg to various perturbations as well as its improvements in diversification of energy sources that reduced the dependence on fossil fuel in favor of bioenergy.

4.6 Concluding Opinion

This chapter aimed to provide an overview of how to evaluate circularity at the meso level, particularly in EIPs. This study showed how EIPs can be identified, how the companies within an EIP interact with each other as well as how the EIPs are managed and relates to the local community. After defining EIPs in all their important elements according to the most relevant literature in the field, the UNIDO EIP toolkit developed to support the implementation of EIPs in existing and new industrial parks were briefly introduced as well as the toolkit's application in evaluating the environmental, economic and social performances of the two EIPs (The IZ NÖ-Süd EIP in Austria and The Ulsan Mipo and Onsan in South Korea). The National EIPs evaluation standard system developed by the Chinese government has been presented through its current indicators along with some of the most significant updates compared to the previous versions of the evaluation standard system. The updates regard the inclusion of indicators measuring the industrial symbiosis category that is a central aspect in an EIP.

Whereas the UNIDO EIP toolkit serves more as a checklist of best practices for an EIP and helps set them apart from normal industrial parks, the Chinese National Evaluation Standard System for EIPs provides indicators to measure the performance of EIPs. In particular, the Austrian Case shows that the EIP meets the imperatives of sustainable development in industrial activities by delivering a wide range of benefits (not limited to economic benefits only) that contribute to creating a more favorable context with the local community and a better relationship with the natural environment. In this view, echoing Cohen-Rosenthal [36], eco-industrial development presents an archway for a better future where companies aim for

---

[10] Kalundborg Symbiosis: Effective industrial symbiosis https://www.ellenmacarthurfoundation.org/case-studies/effective-industrial-symbiosis.

[11] Kalundborg Symbiosis: six decades of a circular approach to production https://circulareconomy.europa.eu/platform/en/good-practices/kalundborg-symbiosis-six-decades-circular-approach-production.

continuous improvement of their economic, environmental and social performances.

References

1. Bocken NM, De Pauw I, Bakker C et al (2016) Product design and business model strategies for a circular economy. J Ind Prod Eng 33(5):308–320
2. Ghisellini P, Ji X, Liu G et al (2018) Evaluating the transition towards cleaner production in the construction and demolition sector of China: a review. J Clean Prod 195:418–434
3. Elia V, Gnoni MG, Tornese F (2017) Measuring circular economy strategies through index methods: a critical analysis. J Clean Prod 142:2741–2751
4. Winkler H (2011) Closed-loop production systems-a sustainable supply chain approach. CIRP J Manuf Sci Technol 4(3):243–246
5. Frosch RA (1992) Industrial ecology: a philosophical introduction. Proc Natl Acad Sci 89(3):800–803
6. Frosch RA, Gallopoulos NE (1989) Strategies for manufacturing. Sci Am 261(3):144–153
7. Chertow MR (2000) Industrial symbiosis: literature and taxonomy. Annu Rev Energy Env 25(1):313–337
8. Liu Z, Adams M, Wen Z et al (2017) Eco-industrial development around the globe: recent progress and continuing challenges. Resour Conserv Recycl 127:A1–A2
9. Le Tellier M, Berrah L, Stutz B et al (2019) Towards sustainable business parks: a literature review and a systemic model. J Clean Prod 216:129–138
10. Ghisellini P, Cialani C, Ulgiati S (2016) A review on circular economy: the expected transition to a balanced interplay of environmental and economic systems. J Clean Prod 114:11–32
11. Su B, Heshmati A, Geng Y et al (2013) A review of the circular economy in China: moving from rhetoric to implementation. J Clean Prod 42:215–227
12. Cappellaro F, Cutaia L, Innella C et al (2019) Investigating CirculaR Economy Urban Practices in Centocelle, Rome District. Environ Eng Manag J 18(10)
13. Van Berkel R, Fujita T, Hashimoto S et al (2009) Industrial and urban symbiosis in Japan: analysis of the eco-town program 1997–2006. J Environ Manag 90(3):1544–1556
14. Côté R, Hall J (1995) Industrial parks as ecosystems. J Clean Prod 3(1–2):41–46
15. Côté RP, Cohen-Rosenthal E (1998) Designing eco-industrial parks: a synthesis of some experiences. J Clean Prod 6(3–4):181–188
16. Lowe E, Moran S, Holmes D (1995) Report for the US protection agency. Indigo Development International, Oakland (CA)
17. Chertow MR (2007) "uncovering" industrial symbiosis. J Ind Ecol 11(1):11–30
18. Susur E, Hidalgo A, Chiaroni D (2019) A strategic niche management perspective on transitions to eco-industrial park development: a systematic review of case studies. Resour Conserv Recycl 140:338–359
19. Shi H, Tian J, Chen L et al (2012) China's quest for eco-industrial parks, part i. J Ind Ecol 16(1):8
20. Halog A, Balanay R, Anieke S et al (2021) Circular economy across Australia: taking stock of progress and lessons. Circular economy and sustainability, pp 1–19
21. Valenzuela-Venegas G, Salgado JC, Díaz-Alvarado FA (2016) Sustainability indicators for the assessment of eco-industrial parks: classification and criteria for selection. J Clean Prod 133:99–116
22. Chertow M (2004) Industrial symbiosis. In: Encyclopedia of energy, Elsevier, San Diego
23. Zhu Q, Lowe EA, Ya Wei et al (2007) Industrial symbiosis in China: a case study of the Guitang Group. J Ind Ecol 11(1):31–42
24. Gómez AMM, González FA, Bárcena MM (2018) Smart eco-industrial parks: a circular economy implementation based on industrial metabolism. Resour Conserv Recycl 135:58–69
25. Bain A, Shenoy M, Ashton W et al (2010) Industrial symbiosis and waste recovery in an Indian industrial area. Resour Conserv Recycl 54(12):1278–1287
26. Graedel TE, Allenby BR (1995) Industrial ecology. Prentice Hall, Englewood Cliffs, NJ, USA
27. Dong H, Geng Y, Xi F et al (2013) Carbon footprint evaluation at industrial park level: a hybrid life cycle assessment approach. Energy Policy 57:298–307
28. Geng Y, Liu Z, Xue B et al (2014) Emergy-based assessment on industrial symbiosis: a case of Shenyang Economic and Technological Development Zone. Environ Sci Poll Res 21(23):13,572–13,587
29. Chertow M, Gordon M, Hirsch P et al (2019) Industrial symbiosis potential and urban infrastructure capacity in Mysuru, India. Environ Res Lett 14(7):075,003
30. van Beers D, Flammini A, Meylan FD et al (2019) Lessons learned from the application of the UNIDO eco-industrial park toolbox in Viet Nam and other countries. Sustainability 11(17):4687
31. Behera SK, Kim JH, Lee SY et al (2012) Evolution of 'designed' industrial symbiosis networks in the Ulsan Eco-industrial Park: 'research and development into business' as the enabling framework. J Clean Prod 29:103–112
32. Turken N, Geda A (2020) Supply chain implications of industrial symbiosis: a review and avenues for future research. Resour Conserv Recycl 161(104):974
33. Huang B, Yong G, Zhao J et al (2019) Review of the development of China's Eco-industrial Park standard system. Resour Conserv Recycl 140:137–144
34. Jacobsen NB (2006) Industrial symbiosis in Kalundborg, Denmark: a quantitative assessment of economic and environmental aspects. J Ind Ecol 10(1–2):239–255
35. Chertow M, Park J (2016) Scholarship and practice in industrial symbiosis: 1989–2014. Taking stock of industrial ecology. Springer, Cham, pp 87–116
36. Cohen-Rosenthal E, Musnikow J (2003) Eco-industrial strategies. Routledge, London

Circularity at Micro Level: A Business Perspective

5

Rashmi Anoop Patil, Sven Kevin van Langen
and Seeram Ramakrishna

Abstract

Circularity assessment at the micro-level (for businesses) is critical for a systemic transition towards a CE as businesses are an integral part of the economy. There are wide-ranging indicators and varying practices and tools for measuring the circularity of businesses due to lack of a standard to monitor and report progress towards the implementation of the CE. Besides, most small and medium businesses lack the know-how or resources to measure their circularity. To cater to such needs, we present a generic methodology and a comprehensive set of indicators for assessing circularity at the micro-level that aligns well with the CE principles and goals. Further, a real-world case study is discussed based on publicly available data (reported by the company), as an example of how companies are trying to become circular in practice. At the end of the chapter, major strategies (along with examples) are recommended that can aid businesses in chartering towards circularity.

R. A. Patil(✉) · S. Ramakrishna(✉)
The Circular Economy Task Force, National University of Singapore, Singapore 117575, Singapore
e-mail: rashmi.anoop33@gmail.com

S. Ramakrishna
e-mail: seeram@nus.edu.sg

S. K. van Langen
UNESCO Chair in Environment, Resources and Sustainable Development (International Ph.D. Programme), Department of Science and Technology, Parthenope University of Naples, 80143 Naples, Italy

Olympia Electronics, Thessaloniki, Greece

S. Ramakrishna
Department of Mechanical Engineering, National University of Singapore, Singapore 117575, Singapore

Centre for Nanotechnology and Sustainability (NUSCNS), 2 Engineering Drive 3, Singapore 117576, Singapore

Keywords

Micro-level circularity assessment · Environmental sustainability · Business circularity · Circular economy

5.1 Introduction

Businesses, inclusive of small-medium enterprises (SMEs) and large corporations contribute directly to the growth of the local economies (specifically, improving the gross domestic product (GDP)) by bringing in investments and providing employment opportunities, and thereby collectively impacting global economic development. However, alongside the benefits,

R. A. Patil and S. Ramakrishna (eds.), *Circularity Assessment: Macro to Nano*,
https://doi.org/10.1007/978-981-19-9700-6_5

businesses generate large quantities of waste and consume enormous amounts of resources such as raw materials, water, and energy which has resulted in complex environmental issues [1]. Therefore, businesses play a vital role in realizing the systemic transition to a CE from the conventional linear economy. Such a transition also provides businesses with opportunities to enhance value besides reducing their environmental impact [2] by adopting the following strategies in their operations and management.

1. Improving supply-chain and resources management.
2. Eliminating/minimizing waste through better design.
3. Keeping the products, components, and materials in use for as long as possible.

These strategies not only aid the businesses to become circular but also provide a platform to pursue the financial benefits of this profound shift [3, 4].

In recent years, many global companies have initiated the transition towards circularity to achieve sustainability and also improve their ESG ratings to attract investors. In the case of SMEs, this change is less focused due to the lack of economic resources and technical know-how to transform the business into a circular one [5]. Contrastingly, in the past decade, many start-ups with innovative eco-friendly products/services and circular business models are springing up globally as a result of growing ecological awareness and support of accelerators for sustainable businesses. Therefore, there is an imminent need for a standard circularity assessment tool that can facilitate uniform measurements of businesses of various sizes and applicable across all industrial sectors.

For the sake of clarity in understanding the assessment methodologies and various indicators for each level in the hierarchy of an economic system, we have considered businesses at the micro-level as outlined in the ISO/TC 323 explanatory note. This is unlike most literature that has been including products and components in the micro-segment [2, 6, 7] and such a definition has neither been consistently used nor clearly represented in the prior art [8]. This chapter begins with a discus-sion on the need for circularity assessment at the micro-level. Then, we present a generic methodology for individual businesses to measure their circularity with a general list of indicators. The self-assessment method enables the businesses to view their operations and management through the circularity lens, indirectly aiding them to enhance their environmental, social, and economic sustainability. With this background, Apple Inc.'s case study is provided to discuss how companies are trying to become circular in practice. As a logical conclusion, the chapter recommends practices and strategies that businesses can follow in their pursuit of circularity.

5.2 The Need for Circularity Assessment at Micro-level

As the CE is gaining momentum, businesses (SMEs and large corporations) belonging to various industrial sectors need to prepare and involve in the transition to sustain the competition and understand the associated risks and opportunities. To do so, it is necessary for businesses to

1. initially know how circular they are,
2. identify the various aspects of operation and management where there is scope for achieving circularity,
3. set achievable targets towards becoming circular, and
4. monitor progress/improvements resulting from circular activities.

To implement the aforementioned points there is a need for a self-assessment methodology/tool for measuring the circularity of businesses that includes a comprehensive list of indicators.

In recent times, many micro-level circularity indicators have been proposed by researchers through academic publications [9, 10]. Most of them are either based on materials-flow or environmental impact. Some of them focus more on economic factors, biasing the measurement towards monetary benefits. A majority of them being single-dimensional indicators (such as recycling, remanufacturing, reusing), researchers emphasize that there is a need for combining such indicators to formulate better ways of measuring the circularity of businesses in practice. To complement this line of thought, lately, some organizations promoting the transition of traditional businesses to circular ones have come up with their proprietary comprehensive tools for measuring business-level circularity such as the Ellen MacArthur Foundation's Circulytics,[1] and the WBCSD's CTIs.[2] The purpose of such tools is to facilitate resource stewardship and decision-making in operational management to engage businesses in transiting towards circularity.

Currently, there is a lack of a global standard for micro-level circularity assessment which results in non-uniform measurements across all business sectors. However, the emerging circularity assessment standard, ISO 59020 framework can be implemented initially by considering the data published by businesses according to the existing sustainability reporting standards such as the Global Reporting Standard—Environmental (300 series)[3] and the IIRC's guidelines for natural capital reporting.[4]

5.3 Circularity Assessment Approach at Micro-level

The key circularity indicators vary from business to business depending on which industry sector it belongs to. For example, the key circularity indicators of a dairy products company are different from those of a telecommunication company. A dairy product company mainly depends on milk supply and resources such as water and energy for processing, packaging materials, and storage facilities. On the other hand, a telecommunication company is dependent on network infrastructure, electronic and electrical equipment, and energy for its operation. Hence, key indicators for 'resource use' in the case of the dairy products company will be the volume of milk and water

consumed/wasted, which is not applicable for a telecommunication company. Furthermore, for a telecommunication company, the environmental impact resulting from network radiation and the e-waste generated is a key indicator that is completely irrelevant to the dairy products company. This diversity poses a challenge to form a common framework for the circularity assessment of the micro-level. However, a generic methodology can be followed by selecting a subset of circularity indicators from a comprehensive set, that is relevant to the business in consideration.

Two major factors influence the circularity of businesses, namely, the internal factors (including operations and waste management) and the external factors (supply-chain). Therefore, for a comprehensive circularity assessment of a business, it is necessary to take into consideration the supply-chain circularity along with the business's circularity performance. However, care should be taken to avoid the overlap of the supply-chain indicators with the business's internal circularity indicators in the assessment process.

To begin with, a company/business can assess its circularity internally that includes operations and waste management (Stage I; Fig. 5.1). This initial assessment provides insight into the consumption and wastage of resources such as water, energy, and raw materials in various processes and their resulting environmental impact. It also gives an idea of how the business can reduce its environmental footprint through waste management strategies. Once this internal assessment is completed, various aspects of operation and management where there is scope for improvement can be identified and necessary changes can be embraced. For instance, in a manufacturing company, the consumption of water and energy could be significantly reduced by adopting smart manufacturing techniques that make use of digital technologies in place of conventional manufacturing methods. Once such necessary changes are implemented, the circularity evaluation should be reiterated to track the progress.

After achieving notable progress (of around 50%) in the circularity of internal operations and management, the supply chain circularity can be addressed (Stage II; Fig. 5.1). As business is dependent on its supply-chain for resources, the

---

[1] Circulytics https://www.ellenmacarthurfoundation.org/resources/apply/circulytics-measuring-circularity.

[2] WBCSD CTI https://www.wbcsd.org/Programs/Circular-Economy/Factor-10/Metrics-Measurement/Circular-transition-indicators.

[3] GRI https://www.globalreporting.org.

[4] IIRC https://integratedreporting.org/.

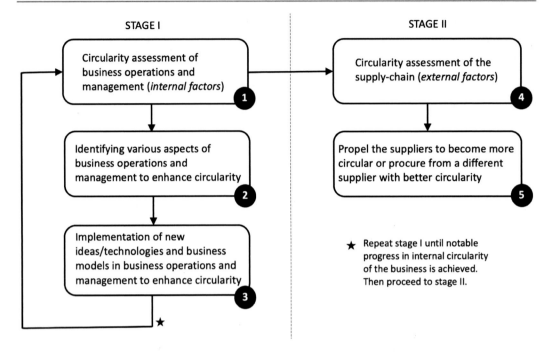

Fig. 5.1 Flow chart showing the stages required for businesses to charter towards circularity. The first stage involves the circularity assessment of internal operations and management and the related implementation of various measures (such as innovative technologies and business models) for enhancing the circularity credentials. The second stage involves assessing and enhancing supply-chain circularity

circularity of the latter becomes critical in achieving the overall circularity at the micro-level. Hence, it is important to have a supply-chain which provides renewable/recycled resources and reduces its carbon footprint in the process [11].

5.4   Indicators for Micro-level Circularity

For a multi-dimensional circularity assessment of a business, it is necessary to measure the circularity of various aspects of the business. Each aspect of business management involves multiple circularity indicators and such quantifiable data expressed in standard form collectively provide the required information for the overall circularity assessment of the business. In this section, we have presented a comprehensive list of indicators categorized based on different aspects of business management as shown in Fig. 5.2. As mentioned earlier, a relevant subset of these indicators can be selected for any business under consideration.

5.4.1   Operations Management

This represents the administrative part of a business that strives to improve the efficiency of the business processes as well as maximize profits. For a business transiting towards circularity, alongside these responsibilities, the operations management should also emphasize on (i) resource efficiency and (ii) eco-design of products/services and processes. These strategies will not only reduce the environmental impact but also aid in achieving economic growth as well as a positive impact on the society.

1. **Resource efficiency:** This strategy aims to use fewer resources for products/services and processes without affecting the quality and quantity of products/services and/or speed of processes. It can be implemented by reducing the resource input and improving resource utilization through redesign, reuse, and remanufacture [12–15]. The following indicators can facilitate enhanced resource efficiency:

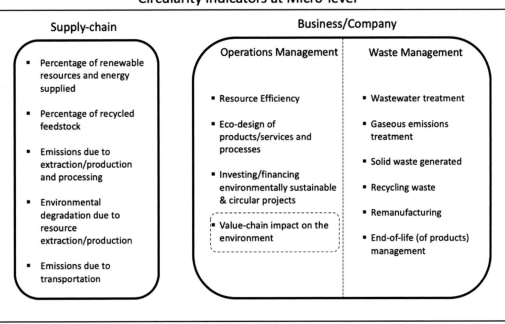

Fig. 5.2 Schematic representation of a comprehensive list of categories of circularity indicators belonging to various aspects of internal operations and management of businesses, and their supply-chains. For a business being considered for circularity assessment, the set of relevant circularity indicators can be chosen from this list

- The ratio of amount of raw materials consumed to that of materials wasted,
- The ratio of the volume of freshwater consumed to the volume of wastewater let out, and
- The ratio of the total energy consumption to energy wastage

2. **Ecodesign of products/services and processes:** Linear businesses have always been about increasing the production rate and profits and so the design philosophy. Contrastingly, in a CE, the design philosophy follows its core principles. Products/processes need to be designed to be reused/remanufactured/recycled (emphasizing ease in disassembly, in terms of process and time required) and to eliminate the use of hazardous materials by finding eco-friendly alternatives [16–18]. The following indicators can provide pointers to adapt changes in redesigning products/services and processes for circularity.

- Feasibility of disassembly and a quick disassembly rate (preferably in minutes) of products for reuse/remanufacture/recycle
- Number and quantities of rare earth and hazardous materials used
- Energy efficiency of products/services and processes

3. **Investing in green projects:** To achieve 100% circularity through becoming carbon-neutral, businesses must invest in green projects such as the generation of renewable energy, afforestation, freshwater replenishment/harvesting programs, and restoring wildlife habitats. These projects help in decreasing the carbon footprint resulting from various business activities [19, 20], and the following indicators need to be evaluated to ensure such a pursuit.

- Investment in green projects (say, in terms of dollars or euros) and

- the reduction in the total carbon footprint of the business value chain (measured in terms of CO_2 eq.).

5.4.2 Waste Management

Although waste management has always been a part of traditional linear businesses, it is mostly in place to abide by rules and regulations and not majorly for reducing or treating waste and environmental protection. Hence, most waste management strategies are not well developed or implemented. CE promotes designing out waste as a principle. So, in the context of CE, the main purpose of waste management is to minimize waste by following various strategies [21, 22]. Such strategies and their corresponding indicators are discussed below.

1. **Wastewater:** Hazardous wastewater generated by businesses pollute the water resources and subsequently affects the biome depending on the water bodies. Therefore, the wastewater has to be treated to remove the toxins before being released into the environment. However, businesses transiting towards circularity also need to recycle and reuse the spent water to reduce water wastage and further exploitation of freshwater resources [23, 24]. Hence, measuring the following indicators can aid in reducing freshwater consumption and pollution of water resources:

   - Percentage of treated water reused and
   - Pollution/depletion of water resources

2. **Emissions:** It is common knowledge that there is an emission of greenhouse gases from manufacturing industries even after the gaseous emissions undergo treatment. But, the fact that every business is responsible for some kind of emission (either due to fossil fuel-based energy usage or consumption of resources with a high carbon footprint) goes unnoticed. Hence, every business should measure

   - the total emissions (due to processes, materials consumption, and energy consumption), measured in terms of tons of CO_2 eq. [14, 25] and

   - the emission of individual hazardous gases.

3. **Solid waste:** Many industries generate different types of solid waste in their general operations or in manufacturing. The primary motive of CE is to reduce waste generation and convert waste matter into resources [26]. Therefore, the following measurements become important in adopting circularity-

   - Tons of organic and inorganic waste generated,
   - Percentage of organic waste decomposed, and
   - Percentage of inorganic waste upcycled/recycled.

4. **Recycling of waste:** Materials recycling has been a well-known and key strategy in waste management for years, even before the advent of CE. However, from a CE perspective, it is considered as the least sustainable option compared to other strategies such as reduction and reuse of raw materials and hence, it represents the outermost loop in the Ellen MacArthur Foundation's butterfly model of CE [3]. The CE related recycling indicators [27] should measure two aspects namely,

   - the recycling and valorization efficiency (in percentage or ratio) and
   - the environmental impact of recycling (measured as CO_2 eq. emissions).

5. **Remanufacturing:** Although this is a matured strategy in a few industries, remanufacturing is comparatively new and predominantly CE-driven since its purpose is to extend the useful life of products/components. Remanufacturing includes refurbishment, reconditioning, and repurposing used products into new ones (either the same or different) by using components or parts from the older products. Hence, this approach is both environmentally friendly and economically beneficial. However, the economic feasibility of remanufacturing needs to be pre-calculated considering the buyback and process costs, and market value. The remanufacturing indicators [28, 29] should measure two aspects namely,

- the percentage of parts/components used for remanufacturing and
- the reduction in the environmental impact due to remanufacturing (measured in terms of CO_2 eq. emissions).

6. **End-of-life (EoL) management:** It is another CE strategy which aims at taking back products at the end of their life, to recover and reuse components and materials. This is becoming popular in recent years as a result of extended producer responsibility legislation being introduced in many countries across the world. Measuring the following indicators can aid in EoL management:

- amount of waste diverted from landfills (measured in tons), and
- the environmental and economic impacts of the collection, disassembly, and recovery of products.

5.4.3 Supply-Chain Management

The supply-chain of a business includes the entire network of producers, manufactures, wholesalers, and logistics that provide the necessary resources such as raw materials and other manufacturing components/equipment, water, and energy for the smooth business operations. Managing the supply chain and planning to drive the suppliers towards circularity is critical in achieving the circular goals of a business. For this, it is essential to measure a few supply-chain indicators yearly and urge the suppliers to enhance their circular performance by gradually modifying the procurement and production processes [30–32]. The following are a few general indicators to be measured for any supplier.

- Ratio of renewable to non-renewable material resources supplied
- Ratio of renewable to non-renewable energy supplied
- Percentage of recycled material resources supplied
- Emissions due to extraction/production and processing

- Environmental degradation/pollution due to resource extraction/production
- Emissions due to transportation.

5.5 Case Study: Apple Inc.

Apple Inc., a trillion-dollar company with a global presence, is a designer, manufacturer, and retailer of mobile communication and media devices, personal computers, portable digital music players, and their products related software, services, accessories, networking solutions, and third-party digital content and applications. Such a company is responsible for a significant impact on the environment. Therefore, they have been putting in efforts to improve their environmental performance for more than a decade, and in doing so, they have adopted the CE intending to become environmentally sustainable. Recently, Apple announced its commitment to become carbon neutral across its entire business, inclusive of the supply-chain.

The intriguing question to be considered here is- 'What's Apple doing to become more circular?' The environmental sustainability team at Apple Inc. is working on becoming circular on various fronts as outlined in Table 5.1, and these efforts towards circularity are as discussed below.

1. **Energy:** With continuous efforts, all of Apple's corporate facilities (inclusive of offices, stores, and data centers worldwide) are powered by renewable energy since 2018. This was made possible through investments in renewable energy projects, such as solar arrays, wind farms, and bio-gas fuel cells (created and owned by Apple), that provide over 80% of the electricity required for its corporate facilities. Other renovation efforts such as LED lighting and upgraded heating and air-conditioning systems have led to further improvements in energy efficiency. These efforts have resulted in lowering electricity needs by nearly one-fifth and saving the company 27 M\$ in 2019. The emissions from the facilities has reduced to $<1\%$ of the total carbon footprint ($>50\%$ decrease in the past decade), preventing more

Table 5.1 Environmental performance indicators relating to Apple's global facilities [33]

Category (Unit)	Key performance indicators
Greenhouse Gas Emissions (metric tons CO_2e)	Emissions from fuel combustion for product transport and heating (A)
	Emissions from Apple's use of electricity (B)
	Emissions from employee commute and business travel (C)
	Total emissions (A + B + C)
Energy Use (million kWh, million btu)	Electricity consumption
	Natural gas consumption
Energy Efficiency (kWh, million btu)	Electricity (kWh) saved as a result of energy efficiency measures
	Natural gas (btu) saved as a result of energy efficiency measures
Renewable Energy (%, metric tons CO_2e)	Renewable energy sourcing (%)
	Emissions avoided as a result of renewable energy sourcing
Water Use (million gallons)	Total usage from offices, data centers and retails
Waste Generation (pounds, %)	Landfilled
	Recycled
	Composted
	Hazardous waste
	Waste to energy
	Landfill diversion rate (%)

than 2 million metric tons of greenhouse gases from entering the atmosphere in the past decade.

Manufacturing processes contribute to more than three-quarters of the company's total carbon footprint and most of it is due to electricity consumption. So, Apple is liaising with the parts suppliers in reducing their energy usage and transiting to renewable energy. Audits and assessment of suppliers' facilities worldwide with a cost-benefit analysis have resulted in significant energy efficiency improvements. Also, Apple has urged its parts suppliers to invest in renewable energy projects and opt for clean energy from utility providers instead of energy from fossil fuels. These endeavors makeup to 4 GW from renewable energy sources and have reduced approximately one-third of its manufacturing carbon footprint. Apple now has commitments from all of its 73 suppliers to use 100% renewable energy for the production of parts by 2030 which is equivalent to nearly 8 GW of power

required for manufacturing. Once this target is reached, it will cut down over 14.3 million metric tons of CO_2e emission annually.

Apple has also been working on improving the energy efficiency of its devices. Over the past 11 years, it has reduced the average energy needed for a product's functioning by 73%.

Another way in which Apple is trying to reduce fossil fuel consumption is by reducing commuting. The total transportation emissions from business fleet vehicles, employee commute, and business travel are reduced by enabling telecommuting (working from home), offering mass transport facilities, and providing electric vehicle charging ports (powered by renewable energy sources)/ bicycles on campus.

2. **Resources:** In 2017, Apple announced its commitment to a closed-loop supply chain in which recycled and renewable materials are sourced, used efficiently for a long time, and replenished back into the supply chain. They have identified a set of materials (such as alu-

Fig. 5.3 Schematic representation of strategies that can be adopted by businesses transiting towards circularity. The strategies are in alignment with the three principles of the CE (design out waste and pollution, keep products and materials in use, and regenerate natural systems)

minum, cobalt, copper, glass, paper, plastics, stainless steel, tin, tungsten, and rare earth elements) used in the manufacturing and packaging of Apple products, to minimize their usage (by redesigning processes and products) and recover them back at the products' end-of-life. They have also strived to eliminate harmful materials such as lead, mercury, beryllium from their processes. To improve the recovery rate of materials and components from used devices, they have employed robotic systems that disassemble the devices efficiently. This is complemented by encouraging consumers with incentives to give back end-of-life devices for recycling. Apple also takes back its devices and refurbishes them like new. Such Apple Certified Refurbished products are recirculated in the secondary market until they reach end-of-life. When it comes

to packaging, the company uses responsibly sourced virgin fiber and recycled paper for product packaging as an initiative to protect forests.

3. **Water:** Apple has come up with the Clean Water Program to help its suppliers (components manufacturers) to conserve water and prevent water pollution. They are working together to reduce water usage, especially for the water-intensive manufacturing processes and in water-stressed regions. Besides the reduction in water consumption at corporate facilities, Apple is opting for alternatives to freshwater such as recycled water, reclaimed water, and harvested rainwater. They are also investing in projects to restore natural water resources inclusive of the groundwater table.

By implementing the aforementioned strategies, Apple Inc. is not only reducing its environmental footprint but also making a positive impact on society as well as reaping the financial benefits of becoming environmentally sustainable. Nevertheless, there is a long way ahead towards achieving a circular and sustainable business.

5.6 Concluding Opinion: Key Strategies and Examples for Practitioners of Business Circularity

The Ellen MacArthur Foundation's Circular Economy System Diagram [3] popularly known as the butterfly model illustrates the 6R strategy for achieving circularity. In this section, we discuss the key strategies [34] shown in Fig. 5.3 with relevant examples that both emerging and established businesses can adopt to become circular. These strategies are in alignment with the core principles of the CE.

1. **REthink and REdesign the way businesses are created:** Businesses need to become circular from the conception stage. Whether it is an established business or a start-up, implementing innovative ideas and new business models will aid this purpose.

   For example, Lehigh Technologies,[5] an SME, introduced an unconventional novel idea to address rubber wastage. The company produces an engineered material called micronized rubber powder from end-of-life tires and post-industrial rubber (that are generally disposed of in landfills) using a proprietary cryogenic turbo mill technology. The rubber powder has been a suitable material for several consumer and industrial applications, including tires, plastics, asphalt, and construction materials serving as an alternative to virgin rubber and fossil fuel-based materials. Such a creative business idea enables closing the loop of highly consumed products such as tires.

2. **REduce material consumption and eliminate wastage:** This is the most important strategy that every business belonging to any industrial sector should adopt. It can be implemented by embracing smart industrial technologies such as artificial intelligence, robotics, internet-of-things (IoT), along with innovative industrial processes, minimal design philosophy and packaging to reduce energy, water, and raw material consumption.

   For example, DyeCoo[6] has come up with new industrial technology to dye fabrics without using water, thereby eliminating water wastage and wastewater treatment (a huge drawback of the conventional textile industry). Compared to the conventional dyeing process, the DyeCoo process is energy efficient and does not require any process chemicals except for pure dyes. Also, the technology uses reclaimed CO_2 in a highly pressurized supercritical state as the dyeing medium, in a closed-loop process (without emitting CO_2 into the environment) and is highly resource-efficient.

3. **REfurbishing and REmanufacturing:** Every business should consider taking back its products for refurbishing/remanufacturing and building a secondary product market. This requires designing products for remanufacturing, establishing collection channels, accumulating know-how to establish remanufacturing processes, and controlling product quality to stimulate demand for remanufactured products. This strategy not only increases the longevity of products' life but also improves the overall circularity of the business.

   For example, Fuji Xerox, Ricoh, and Canon, which collectively hold nearly 90% of Japan's photocopier market, remanufacture their photocopiers. In the case of Fuji Xerox, there is no distinction between completely new and remanufactured photocopiers, and any machine may contain used components. Contrastingly, Ricoh and Canon separate their completely new photocopiers from remanufactured ones.

[5] Lehigh Technologies https://lehightechnologies.com/.

[6] DyeCoo http://www.dyecoo.com/.

Another example of the implementation of this strategy is IKEA that recently opened new stores in Sweden for exclusively refurbished furniture as a part of their CE initiative.

4. **REdistribute for REuse:** This includes resale or donation of used products in good condition from the first owner (who may no longer need the item) to someone who may use it. Such secondary usage of products facilitates extending the life of the products.

   For example, many e-commerce retailers allow users to sell preowned goods in a usable condition aiding in extending products' life. This new concept is becoming popular among consumers, especially for expensive products.

5. **REcycle:** CE provides opportunities for new businesses/products, with innovative methods for recycling industrial and post-consumer wastes. Such businesses are essential to increase the recycling rate and efficiencies. This will consequently lead to an improved secondary material market that has the potential to be the supply-chain for businesses.

   For example, Adidas partnered with Parley,[7] has come up with a new collection of products that include high-performance sportswear shoes and jerseys made from upcycling the polyethylene terephthalate (PET) bottles polluting the oceans. Each product contains at least 75% of polyester from recycled marine trash and utilizes less water and fewer chemicals for production. This also has the potential to decouple the business from fossil-fuel based raw materials.

6. **Investing in green projects:** When renewable energy or sustainably sourced water is scarce, investing in green projects (such as the generation of renewable energy and rainwater harvesting) which match the resource needs of the business makes it more circular. If circularity is viewed at the molecular level (in the context of carbon emissions), any business activity that is material-wise circular

produces greenhouse gases. Hence, investing in projects such as afforestation or restoring natural habitats will aid in businesses becoming carbon neutral or circular at a molecular level. Recently, many global corporations have begun investing in green projects.

For example, the Coca-Cola company,[8] a huge consumer of water, replenishes 100% of water consumed in the entire range of its bottled products and their production processes, by increasing water usage efficiency in its plants and returning water to the sources through wastewater treatment. The company also engages in diverse, locally focused community water projects with objectives such as providing or improving access to safe water and sanitation and conserving natural water resources.

The other two strategies namely, IS and repair to prolong the products' life (especially technological ones), have been discussed in detail along with relevant case studies in Chaps. 44 and 77 respectively.

References

1. McConnell JR, Edwards R, Kok GL et al (2007) 20th-century industrial black carbon emissions altered arctic climate forcing. Science 317(5843):1381–1384
2. Ghisellini P, Cialani C, Ulgiati S (2016) A review on circular economy: the expected transition to a balanced interplay of environmental and economic systems. J Clean Prod 114:11–32
3. Ellen MacArthur Foundation (2013) Towards the circular economy: economic and business rationale for an accelerated transition
4. Lieder M, Rashid A (2016) Towards circular economy implementation: a comprehensive review in context of manufacturing industry. J Clean Prod 115:36–51
5. Stahel WR (2016) The circular economy. Nature 531(7595):435–438
6. Kirchherr J, Reike D, Hekkert M (2017) Conceptualizing the circular economy: an analysis of 114 definitions. Resour Conserv Recycl 127:221–232
7. Corona B, Shen L, Reike D et al (2019) Towards sustainable development through the circular economy-a review and critical assessment on current circularity metrics. Resour Conserv Recycl 151(104):498

[7] Adidas and Parley https://www.adidas.com.sg/parley.

[8] The Coca-Cola Company https://www.coca-colacompany.com/sustainable-business/water-stewardship.

8. Moraga G, Huysveld S, Mathieux F et al (2019) Circular economy indicators: what do they measure? Resour Conserv Recycl 146:452–461

9. Kristensen HS, Mosgaard MA (2020) A review of micro level indicators for a circular economy-moving away from the three dimensions of sustainability? J Clean Prod 243(118):531

10. Rossi E, Bertassini AC, dos Santos Ferreira C et al (2020) Circular economy indicators for organizations considering sustainability and business models: plastic, textile and electro-electronic cases. J Clean Prod 247(119):137

11. Kazancoglu Y, Kazancoglu I, Sagnak M (2018) A new holistic conceptual framework for green supply chain management performance assessment based on circular economy. J Clean Prod 195:1282–1299

12. Kalliski M, Engell S (2017) Real-time resource efficiency indicators for monitoring and optimization of batch-processing plants. Canad J Chem Eng 95(2):265–280

13. Huysman S, Sala S, Mancini L et al (2015) Toward a systematized framework for resource efficiency indicators. Resour Conserv Recycl 95:68–76

14. Park HS, Behera SK (2014) Methodological aspects of applying eco-efficiency indicators to industrial symbiosis networks. J Clean Prod 64:478–485

15. Tyteca D (1998) Sustainability indicators at the firm level: pollution and resource efficiency as a necessary condition toward sustainability. J Ind Ecol 2(4):61–77

16. Rodrigues VP, Pigosso DC, McAloone TC (2017) Measuring the implementation of ecodesign management practices: a review and consolidation of process-oriented performance indicators. J Clean Prod 156:293–309

17. Cerdan C, Gazulla C, Raugei M et al (2009) Proposal for new quantitative eco-design indicators: a first case study. J Clean Prod 17(18):1638–1643

18. Aoe T (2007) Eco-efficiency and ecodesign in electrical and electronic products. J Clean Prod 15(15):1406–1414

19. Sparkling AE (2012) Cost justification for investing in LEED projects. McNair Scholars Res J 4(1):7

20. Blignaut J, Aronson J, de Groot R (2014) Restoration of natural capital: a key strategy on the path to sustainability. Ecol Eng 65:54–61

21. El Haggar S (2010) Sustainable industrial design and waste management: cradle-to-cradle for sustainable development. Academic, Cambridge

22. Zvolinschi A, Kjelstrup S, Bolland O et al (2007) Exergy sustainability indicators as a tool in industrial ecology. J Ind Ecol 11(4):85–98

23. Liu H, Wang H, Zhou X et al (2019) A comprehensive index for evaluating and enhancing effective wastewater treatment in two industrial parks in China. J Clean Prod 230:854–861

24. Molina-Sánchez E, Leyva-Díaz JC, Cortés-García FJ et al (2018) Proposal of sustainability indicators for the waste management from the paper industry within the circular economy model. Water 10(8):1014

25. Kolokoltsev V, Vdovin K, Mayorova T et al (2017) Ecological indicators in the system of non-financial reporting at industrial enterprises

26. Huysman S, De Schaepmeester J, Ragaert K et al (2017) Performance indicators for a circular economy: a case study on post-industrial plastic waste. Resour Conserv Recycl 120:46–54

27. Di Maio F, Rem PC et al (2015) A robust indicator for promoting circular economy through recycling. J Environ Prot 6(10):1095

28. Golinska-Dawson P, Kosacka M, Werner-Lewandowska K (2018) Sustainability indicators system for remanufacturing. In: Sustainability in remanufacturing operations, Springer, Berlin, pp 93–110

29. Jiang Z, Ding Z, Zhang H et al (2019) Data-driven ecological performance evaluation for remanufacturing process. Energy Convers Manage 198(111):844

30. Schaltegger S, Burritt R, Varsei M et al (2014) Framing sustainability performance of supply chains with multidimensional indicators. Supply Chain Manag: Int J

31. Schaltegger S, Burritt R, Bai C et al (2014) Determining and applying sustainable supplier key performance indicators. Supply Chain Manag: Int J

32. Clift R (2004) Metrics for supply chain sustainability. In: Technological choices for sustainability. Springer, Berlin, pp 239–253

33. Apple Inc (2019) Environmental progress report

34. Dewulf J, Van Langenhove H (2005) Integrating industrial ecology principles into a set of environmental sustainability indicators for technology assessment. Resour Conserv Recycl 43(4):419–432

Circularity at Nano Level: A Product/Service Perspective

6

Rashmi Anoop Patil, Sven Kevin van Langen and Seeram Ramakrishna

Abstract

Implementation of the CE principles at the grass-root level requires closing the loop of products' life cycles, one that warrants extensive rework at every stage of the life cycle from design to end-of-life. Therefore, the circularity assessment of products/services for their entire life cycle becomes critical. In this work, we present the circularity assessment of products/services (considered as the nano level in the economic hierarchy) unlike the prior art and provide a comprehensive metric required for this assessment. The chapter begins with a brief explanation of the need for such an evaluation. Then, continues with a discussion on the state-of-the-art assessment of a product along with a list of various indicators that have been implemented in the circularity measurements at the nano level. This theoretical background is complemented with real-world case studies of Levi's jeans and surgical face masks for a better understanding of the concept. Then, the concept of materials utilization efficiency, a key factor in conceiving and designing circular products/services, is presented as a logical conclusion to the chapter.

R. A. Patil(✉) · S. Ramakrishna(✉)
The Circular Economy Task Force, National University of Singapore, Singapore 117575, Singapore
e-mail: rashmi.anoop33@gmail.com

S. Ramakrishna
e-mail: seeram@nus.edu.sg

S. K. van Langen
UNESCO Chair in Environment, Resources and Sustainable Development (International Ph.D. Programme), Department of Science and Technology, Parthenope University of Naples, 80143 Naples, Italy

Olympia Electronics, Thessaloniki, Greece

S. Ramakrishna
Department of Mechanical Engineering, National University of Singapore, Singapore 117575, Singapore

Centre for Nanotechnology and Sustainability (NUSCNS), 2 Engineering Drive 3, Singapore 117576, Singapore

Keywords

Nano-level · Environmental sustainability · Product circularity · Circular services · Circular economy

6.1 Introduction

Products and services represent the most basic and key components of an economy and are traded in exchange for currency. In the past few decades, advances in science and technology, and the increasing affordability of consumers and businesses have contributed to the exponential growth in the production and consumption of goods. In this process, natural resources and virgin raw materials have been exploited and wasted irresponsibly due to the existing linear economic model [1, 2]. This has led to unintended environmental impacts such as the depletion of natural resources (almost to an irreversible extent), and environmental pollution on both regional and global scales [3, 4]. Therefore, the way products and services are made, used, and disposed of plays a crucial role in transiting to a CE [5].

In recent years, many large corporations are willingly putting efforts into improving product design and services to create value and become more environmentally sustainable. Some of them are even pursuing certifications from organizations such as the Cradle to Cradle Products Innovation Institute and the Ellen MacArthur Foundation. Various digital tools and sets of indicators such as the CET, the MCI [6], and the CEIP have been made available to businesses/reachers/analysts to measure the circularity of commercial products [7]. Startups around the world championing the CE are coming up with eco-friendly products and services, and innovative business models to reduce the environmental impacts of consumables.[1,2,3] However, the eco-friendliness of products and services does not guarantee their circularity. This is why the circularity assessment of products and services that can identify the extent of their circularity and potential avenues for further improvement becomes important.

Through this chapter, we distinguish the circularity assessment of products and services—considered as nano-level in the economic hierar-chy, from the micro-level (businesses), in contrast to the prominent prior art [8–10]. This chapter begins with a discussion on the need for circularity assessment of the nano-level: products and services. Then, we present a background of the state-of-the-art assessments of products at various stages of a product's life-cycle such as design, production, use, and end-of-life as well as the list of indicators used in measuring the different aspects of circularity at the nano-level. With this background, a couple of case studies are illustrated for a better understanding of the product's environmental impact assessment. Since the product circularity directly depends on material circularity, a discussion on the efficient utilization of materials in product design is presented as a logical ending for the chapter.

6.1.1 The Need for Circularity Assessment at Nano-Level

Many global businesses and start-ups are transiting towards circularity for a sustainable business and with new/upcoming policies of governments on environmental sustainability and the rise in conscious consumerism, every product and service is required to become circular. To do so,

1. the design and manufacturing need to adhere to the principles of the CE [5, 11, 12],
2. the useful phase of products have to be extended for as long as possible [11], and
3. at the end-of-life, products have to be collected for recycling [11].

This will ensure the sustainability of both natural ecosystems and the economic system in the long-term. Moreover, in the short-term, circularity will reduce the stress on scarce material resources and decouples economic growth from volatile commodities. From a financial perspective, although transiting to circularity needs initial investments, circular products and services have the potential to be profitable in the long run [2].

Although there are a few organizations and toolkits providing product circularity assessment, there is a need for a standard methodology for uniform evaluation across products that belong to the same category from different manufacturers. For example, the circularity of laptops/computers produced by various manufacturers across the world,

---

[1] The role of startups in accelerating circular economy https://circulareconomyloop.com/the-role-of-startups-in-accelerating-circular-economy/.

[2] Closing the loop: 20 circular startups making useful things out of waste https://www.positive.news/economics/circular-economy-20-startups-making-useful-things-out-of-waste/.

[3] 10 promising circular economy startups to watch in 2021 https://www.eu-startups.com/2021/05/10-promising-circular-economy-startups-to-watch-in-2021/.

or cosmetics of various commercial brands in the market have to be evaluated using a standard methodology for better comparison of their sustainability. Such an assessment should also identify and provide insights to improve the overall circularity of the product.

6.2 Circularity Assessment Approach at Nano-Level

When it comes to the transition of a linear product towards circularity or designing of a new circular product, the focus is on the materials involved, both ingredients and other resources needed in the making. Whether the product is technological or biological, the underlying fundamental principles for the transition are the same. The fundamentals established by the pioneer experts such as Stahel, McDonough, and Braungart (as below) concentrate on the efficient utilization of resources in such a way that they can be restored back into the system.

1. Minimizing the usage of resources and increasing the resource efficiency
2. Slowing down the rate of consumption of resources
3. Closing the resource loops

All the practical strategies in transiting towards circular products such as eco-design, life extension, upcycling of resources, innovative business models, and the indicators used to measure the extent of circularity as well as the progress milestones are all based on these principles. With this background, it can be said that the indicators used for the circularity assessment of a product should measure the extent to which its material components are in a restorative flow, the lifespan or the number of uses compared to similar products (from competitors) [2], the economic value of the product (in particular, the ingredient materials) at its end-of-life, and also its social impact. The assessment of this data provides the overall circularity of a product, say a value between 0 and 1 or 1 and 100, and the necessary information to identify various aspects throughout its life-cycle to enhance the product circularity.

In general, a product that is manufactured using only virgin raw materials dumped in a landfill/incinerated at the end-of-life is considered a linear product [2]. For example, a single-use plastic shopping bag made of petroleum-based virgin plastic disposed of in a landfill is linear, scoring 0 on a scale of 0–1. Contrastingly, a product made of recycled/renewable materials with high recycling efficiency and collected for recycling/component reuse is considered a circular product [2]. For example, a pair of sports shoes made from 100% recycled plastics, used for a long time and collected back for recycling at the end-of-life, scores nearly 1 as the efficiency of recycling/upcycling is taken into account! However, most commercial products which are chartering towards circularity can be placed between 0 and 1, as they are partially circular.

Assessing the circularity of a product initially requires the understanding and analysis of various phases of the product's life-cycle, namely, the design phase, the manufacturing phase, the useful phase, and the end-of-life phase (as shown in Fig. 6.1). For this, it is necessary to answer the following questions:

1. How circular is the product by design?
2. How resource-efficient is the manufacturing process?
3. How long or how many times the product can be used?
4. What is the upcycling/recycling/remanufacturing efficiency of the product?

These will provide an overall perspective of how circular the product is and what steps can be implemented to improve its circularity. Measuring the circularity of a product/service requires data obtained using indicators related to the aforementioned influencers.

6.2.1 Measuring Circularity of a Product and Circularity Indicators

In the past decade, many indicators have been proposed and implemented to assess an individual product (nano-level). Most of them are contributions from academic researchers [13] and a few tools/models involving multiple individual indicators presented by groups of experts from various organizations. These indicators can be broadly classified into (i) individual quantitative

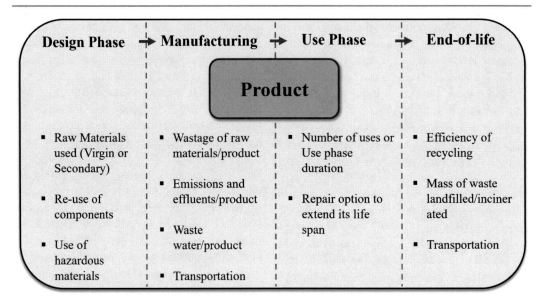

Fig. 6.1 Schematic representation of the factors influencing the circularity of a product at all stages of its life-cycle. Each of the influencers can be quantified using multiple indicators to measure the total circularity of a product considered

indicators and (ii) composite sets of indicators (including indexes, tools, and models) combining multiple individual indicators. These indicators generally focus on measuring one or two aspects of a product's circularity rather than the total circularity throughout its life cycle. To better understand and analyze these indicators, they have been grouped into eight categories based on what they measure, namely, design, resource efficiency, lifetime extension, disassembly, remanufacturing, recycling, end-of-life management, and waste management as shown in Table 6.1.

These indicators either measure how circular a phase in the product's life cycle is concerning its resource/material mass or the economic feasibility of a particular circular process such as disassembly, recycling and remanufacturing. However, there are no indicators measuring the social contributions of a circular product. More importantly, there is a need to combine these indicators into a comprehensive set and close the gaps with new indicators for measuring the total circularity of a product. Besides, attention must be paid to avoid repeated counting of data for multiple indicators. This will ensure accuracy in uptake (for assessing

higher levels of the economic hierarchy) or scaling of circularity information. Lastly, standardization of measuring processes (including the methods protocol and how the results must be expressed) and mandatory regulations for following the standards are quintessential in the evaluation of the nano-level.

6.3 Broadly Applied Methods and Case Studies

Currently, products are majorly being assessed either using MCI [19] or LCA [34] for understanding the circularity and sustainability respectively. Although the two methods seem similar, there are a few major differences. For example, MCI's focus is on product-centric resource efficiency; whereas LCA concentrates mainly on the environmental impact of the product at all stages of its life-cycle. To elaborate, the LCA considers the environmental impact of the raw material extrac-

Table 6.1 Circularity indicators for circularity assessment at the Nano level

Category	Indicator/Index	Description of indicators
Design	Eco-efficiency for ecodesign [43]	A quantitative indicator represented as ratio of the product/service value to the environmental impact of product/service, providing insights for an eco-design.
	Design Methodology for End-of-use Product Value Recovery [14] and End-of-life Indices - Design [15]	The end-of-use strategies such as recycling, remanufacturing, reuse, and disposal are compared based on the cost/revenue approach (disassembly/process costs needed for each strategy) to improve the design for product's better end-of-life performance
	Remanufacturing with Product Profiles [16]	It's an online tool to support product designers to compare different product profiles that are suitable for remanufacturing considering the costs and redesigning potential
Resource efficiency	Eco-cost/Value Ratio [17]	Expresses the resource-efficiency as a ratio of eco-costs and the value of the product
	Value-based Resource Efficiency Indicator [18]	Evaluates the resource-efficiency of a product based on the mass and economic value of the resources used
Lifetime extension	Material Circularity Indicator [19]	Considers the utility of a product by taking into account the lifetime or number of uses of the product to calculate the overall circularity
	Longevity Indicator [20]	Measures the duration for which a material is retained in the product system inclusive of the refurbished/recycled lifetime
	Combination Matrix [21]	It's a combination of the lifetime (in a product system) and the circularity of resources to distinguish products into various categories
Disassembly	Ease of Disassembly Metric [22]	Calculates the time required for the disassembly of a product as well as each of its components
	Effective Disassembly Time[23, 24]	Considers multiples criteria such as disassembly depth, disassembly sequence, and liaison to calculate the disassembly time to isolate a target component
	Disassembly Effort Index [25]	Provides a score based on the efforts and processes to disassemble a product that can be used to calculate the disassembly cost and the disassembly return on investment
Remanufacturing – refurbishing and repurposing	Decision Support Tool for Remanufacturing [26] and Sustainability Indicators System for Remanufacturing [44]	Assess whether remanufacturing is environmentally and economically feasible by calculating the overall process cost and environmental impacts of different remanufacturing yields
	Combination Matrix [20]	The contribution of remanufacturing towards the circularity (a number between one and infinity) and longevity (time) of the product is measured as a proportion of remanufactured products and refurbished lifetime
	Eco-efficient Value Creation Method [27]	Assesses the potential of remanufacturing by combining the costs, market value as well as the environmental costs, and presents a sustainable business strategy matrix to analyze the market prospects of refurbished products/services
	Data-driven Ecological Performance Evaluation [45]	A set of data-driven techniques, such as data envelopment analysis, R clustering and grey relational analysis, are deployed to analyze and evaluate the ecological performance of a remanufacturing process.

(continued)

Table 6.1 (continued)

Category	Indicator/Index	Description of indicators
Recycling	Product Circularity Metric [6]	The ratio of the economic value of recirculated (recycled parts and materials) parts and the economic value of all parts
	Material Circularity Indicator [19]	Considers the mass of recycled content in a product for assessment
	Recycling Index (RI) [28]: (i) Product RI, (ii) Material RI	Considers the mass of the total product, its components and ingredient materials along with a material security index, technology readiness level and simplicity index to deduce how desirable product recycling is.
	Material Reutilization Score (Cradle2cradle Certification) [29]	It represents the combination of the fraction of recycled or rapidly renewable content in a product with the fraction of material that is recyclable and/or biodegradable.
	Circularity Calculator[6]	Calculates the recycled content of a product which includes the recycled content in the original product and the recycled content through closed-loop recycling.
	Circular Economy Index [30]	Ratio of materials value achieved from recycling end-of-life product and the material value of the resources needed for producing the product
	Sustainability Indicators for CE [31]	Takes into account the fraction of recyclable mass in the product and the efficiency of the recycling process
End-of-life management	End-of-life Index [32]	A set of indices that indicate how the product will perform at the end-of-life based on the total cost of each process (disposal, disassembly, recycling, and remanufacturing)
	Product Recovery Multi-criteria Decision Tool [33]	Various end-of-life strategies (reuse, repair, recycling, and remanufacturing) are assessed using relevant economic, environmental, and social indicators
Waste management	Linear Flow Index in MCI [19] and Sustainability Indicators for CE [31]	Takes into account the non-recycled or non-recyclable materials that contribute to waste generation

tion and transportation (inclusive of raw materials, finished products/components, and waste) whereas these factors do not have any weightage in MCI. Another difference between the two is that MCI was designed for mainly technological products, whereas LCA is broadly applicable. However, both methods are dependent on a database created using the MFA.

As discussed earlier, CE is a three-dimensional concept that encompasses environmental, social, and economic aspects. Therefore, for a standard method to assess the circularity of a product, it is necessary to couple the advantages of both MCI and LCA, and also introduce other important factors such as economic competitiveness and social impact (for example, the creation of jobs). The assessment framework must also be universally applicable for products from both technical and biological cycles.

In this section, due to the unavailability of details on MCI analysis and implementation for a specific commercial product [19], we have presented two case studies of commercial products assessed using the LCA method. These case studies clearly show how the assessment is done at every stage of the life-cycle from raw material acquisition to end-of-life using relevant indicators.

6.3.1 Case Study I: Levi Strauss & Co. 501 Jeans Wear

Levi Strauss & Co. (LS&Co.), a pioneer in the apparel industry and a global leader in jeanswear, assessed the environmental impacts of their core set of products using LCA. Their first study was focused on the company's US operations and was further expanded to cover the global scope. During the studies, they analyzed the environmental impacts of a pair of Levi's® 501® jeans, a pair of Levi's® Women's jeans, and a pair of Dockers® Signature Khakis.

To understand how the LCA is applied to a commercial product, the case of the analysis of Levi's® 501 jeans at every stage of its life cycle is presented. The life cycle of Levi's® 501 jeans includes 6 phases (as shown in Fig. 6.2) [35, 36] and is listed below along with the key aspects of each phase.

1. **Raw material production:** Natural fibers agriculture (such as cotton which is 91% of the raw material), synthetic materials manufacturing, regeneration of fibers, extraction of metals (for buttons and zippers), materials derived from livestock (such as leather)
2. **Intermediate production of fabric:** Extrusion, spinning, weaving, dyeing, finishing, and sundries production (inclusive of molding and forming)
3. **Apparel production** (inclusive of packaging): Garment assembly (involving cutting, sewing, sundries application), garment finishing, and garment dyeing
4. **Transportation and distribution:** Transportation of raw materials and intermediate products, distribution of finished apparel to the retail stores, and waste disposal
5. **Use phase:** Wearing, washing, drying, ironing, and repairing
6. **End-of-life/Recycling:** Landfilling, incineration, biodegradation, and recycling.

Every phase of the life cycle consumes energy, water, and chemicals, and generates waste impacting the environment. Therefore, the amount of energy, water, and chemicals consumed and their environmental impact in terms of the carbon footprint are measured. Energy consumption for electricity, heating, and transport is measured in terms of the amount of fossil and renewable resources required for its generation. Measuring water depletion includes consumption of water from freshwater sources and pollution of water resources by the effluents. Regarding chemical footprint in the product's life cycle, each phase requires different types of chemicals, spanning from fertilizers and pesticides for cotton cultivation, process chemicals such as dyes and detergents, and finishing chemicals, to plastics used in packaging. Hence, each of the chemicals used has to be taken into account with the respective usage quantities and the related environmental impact.

The measured indicators are further segregated into various impact categories to analyze the apparel's impact on the environment throughout its life cycle. Then, the respective measured values are aggregated to find the total impact of the product in a particular category. The

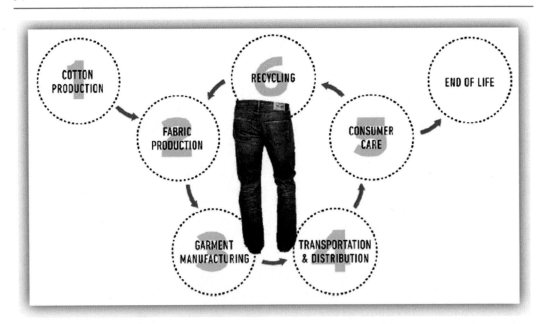

Fig. 6.2 Schematic illustration of the various phases of the life cycle of a pair of Levi's® 501 jeans and the equivalent environmental impact caused in its entire life-cycle. Reproduced from [35]; Copyright, 2017 Elsevier

environmental impact categories with their total measured values of one pair of Levi's® 501 jeans obtained from the LCA are listed in Table 6.2.

The LCA revealed that the major contributor to environmental degradation is the consumption of (i) water and (ii) non-renewable energy resources, for the production of cotton and consumer care (inclusive of washing frequency and drying across the globe) respectively. The other phases of the life cycle have a comparatively lesser impact on the environment.

To address these issues, LS&Co. has devised and implemented programs [36] such as (i) Water<Less® committing to reduce 50% of water being used by 2025, (ii) Better Cotton Initiative to source sustainably produced cotton, (iii) Wellthread to manufacture sustainable alternate raw materials, (iv) Responsible Sourcing inclusive of screening chemicals for Zero Discharge of Hazardous Chemicals (ZDHC) standards, adapting sustainable energy sources, and most importantly, educating consumers about their role in lowering the carbon footprint of their products.

6.3.2 Case Study II Single-Use Surgical Face Mask

The COVID-19 pandemic, being an airborne infection spreading rapidly, has compelled people around the world to wear face masks as physical protection from the transmission of the infection [38, 39]. This has resulted in increased consumption of face masks, especially the single-use surgical type, as it needs zero maintenance and is economical. Consequently, billions of such masks are being manufactured, used, and disposed of; creating a huge environmental impact due to the resulting waste, compelling researchers across the world to study and assess the impacts.

This case study presents the environmental impact of one functional unit (defined as the average number of masks needed per person for one month i.e. 30 masks at an average use of 1 mask/day) of single-use surgical face masks. The assessment was carried out using the LCA and the impacts have been represented in nine different categories. Since this study has been conducted in Singapore, the parameters considered in the analysis, such as the end-of-life treatment and transportation, are specific to this case.

Table 6.2 Categories of environmental impact obtained from the LCA of one pair of Levi's® 501 jeans, measured and reported by Levi Strauss & Co. [36]

Category	Description	Measured value (Unit)
Climate change	Total Greenhouse Gas Emissions fueling the global warming	33.4 (kg CO_2-e)
Water consumption	Net fresh water drawn from the environment minus the water replaced at the same quality or better	3781 (liters)
Eutrophication	Oxygen depletion in water bodies due to the release of nitrogen and phosphorous (from chemicals such as fertilizers, pesticides)	48.9 (g PO_4-e)
Abiotic depletion	Depletion of non-renewable resources that includes fossil fuels, metals and minerals	29.1 (mg Sb-e)
Land occupation	Total land occupied to support the product system	12 (m^2/yr)

Table 6.3 Environmental impact categories obtained from the LCA of one functional unit of single-use Surgical Face Mask [37]

Category	Description	Measured value (Unit)
Climate change (kg CO_2-eq)	Total Greenhouse Gas Emissions fueling the global warming	0.580 (kg CO_2-eq)
Fossil fuel depletion	The extraction fossil fuels (includes all types of fossil resources)	0.308 (kg oil-eq)
Metal depletion	The extraction of virgin metals (aluminum)	0.045 (kg Fe-eq)
Water depletion	Net fresh water drawn from the environment minus the water replaced at the same quality or better	0.006 (m^3 Water eq)
Freshwater ecotoxicity	The impact of the emissions of toxic substances on freshwater ecosystems	0.033 (kg 1,4-DCB-eq)
Freshwater eutrophication	Oxygen depletion in freshwater bodies due to the release of chemicals containing nitrates and phosphates adversely affecting the aquatic flora and fauna	−0.00012 (kg PO_4-eq)
Marine ecotoxicity	The impact of the emissions of toxic substances on marine ecosystems	0.029 (kg 1,4-DCB-eq)
Marine eutrophication	Oxygen depletion in marine water due to the release of chemicals containing nitrates and phosphates adversely affecting the biota	0.0001 (kg NO_4-eq)
Human toxicity	The potential harm on the human body by a unit of chemical released into the environment (because of the inherent toxicity of a compound and its potential dose)	0.034 (kg 1,4- DCB-eq)
Waste generated	Waste generated due to the disposal of the product at the end-of-life and excludes generated from raw material production	0.004 (kg)

Before discussing the LCA of the mask, its design/construction and end-of-life treatment need to be understood. The single-use surgical face mask is a 3-layered sandwich of melt-blown polypropylene between 2 layers of spun-bond PP with an aluminum nose adapter and polyurethane earloops.[7] These face masks are considered to be contaminated once worn and are either recycled or disposed of. Due to the absence of mask recycling facilities in Singapore, the used masks are incinerated along with the general municipal waste, and eventually, the residue is landfilled.

The study considered all the life-cycle phases from raw material acquisition to end-of-life as shown in Fig. 6.3, and the assessment process of each stage is explained below.

1. **Raw material acquisition**: An inventory of the emission factors of the feedstock required for the production and packaging of the masks is created to measure the cumulative emission.

2. **Production**: The electricity and water usage for the production and packaging processes

---

[7] Design of a 3-ply mask https://www.covid19preventiongear.com/blog/how-does-a-3-ply-mask-work.

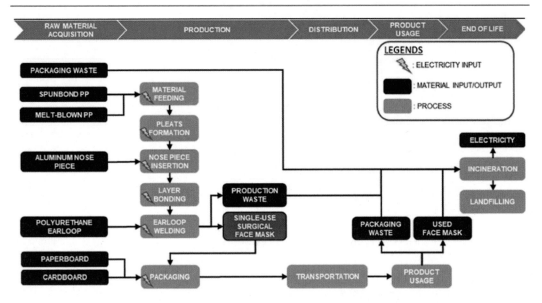

Fig. 6.3 Schematic illustration of the process map including all the stages of the life-cycle of single-use surgical face masks in Singapore. The life cycle of the single-use mask includes the materials input/output, electricity consumption for production, and other processes in the life-cycle contributing to the environmental impact. Reproduced from [37]; Copyright, 2021 Elsevier

are accounted for and their equivalent carbon emissions (in Singapore) were calculated.

3. **Distribution**: The fossil fuels used in transportation are taken into account to calculate the carbon emissions.

4. **Product Usage**: One functional unit (FU) of masks is defined as the consumption of masks for a person in a month (taken as 31 days) with the duration of usage per day assumed to be less than 12 h. Therefore, 1 FU is considered equivalent to 31 single-use surgical face masks and is used to calculate the consumption.

5. **End-of-life**: The amount of waste generated (by weight) after use was calculated and the emissions in treating the waste were taken into account.

The emissions measured were further categorized into major environmental impacts and a cumulative value for each category was obtained. These categories with their respective measured values have been listed in the Table 6.3.

These measured values may seem insignificant when perceived for 1 FU, however, they are quite impactful when scaled to the consumption rate of a large population over a long period. Also, it was identified during the study that raw material acquisition is the highest contributor to the carbon emissions and pollution of water bodies. Therefore, an efficient recycling process could lower the long-term environmental impact of single-use surgical face masks.

6.4 Concluding Opinion: Material Efficiency for Product Circularity

This chapter gives an insight into how most products are currently assessed to understand their environmental impacts. However, circularity assessment does not limit to environmental impacts and extends to social and economic impacts as well, as discussed in Chap. 1. More importantly, the purpose of circularity assessment of products is to slowly close the resource loops using strategies such as redesigning products and introducing new business models [5, 12, 40]. Designing a circular product or refashioning an already existing commercial product into a circular one requires the implementation of multiple strategies.

As we have discussed various circular business models such as extending the lifetime of a product through repairing and IS, we consider design strategies that augur well with the business models. Materials play a vital role when it comes to designing circular products [41, 42]. This means that at the conceptual stage of a product, the ingredient materials chosen should enable circularity and the product design should allow efficient recycling of the used materials. Therefore, the choice of materials should be such that it-

1. allow usage of secondary materials [35] and less material by weight for an efficient product design,
2. is non-hazardous to the environment and the biota,
3. has a low carbon footprint (emissions associated with producing and processing of materials),
4. can be efficiently and economically recycled,
5. promotes the reusability of materials in the form of durable components (for example, hard disks and other electronic components).

Following these strategies in designing products at the nano systemic level can also enable achieving the goal of closing the resource loops at a macro level.

References

1. Ramkumar S, Kraanen F, Plomp R et al (2018) Linear Risks (Joint project between Circle Economy, PGGM, KPMG, EBRD, and WBCSD)
2. Ellen MacArthur Foundation (2013) Towards the circular economy: economic and business rationale for an accelerated transition
3. Wit Mde, Hoogzaad J, Daniels, Cvon (2020) The circularity gap report 2020, circle economy
4. IRP (2017) Assessing global resource use: a systems approach to resource efficiency and pollution reduction; compiled by international resource panel. United Nations Environment Programme, Nairobi, Kenya
5. Mestre A, Cooper T (2017) Circular product design. a multiple loops life cycle design approach for the circular economy. Des J 20:S1620–S1635
6. Linder M, Sarasini S, van Loon P (2017) A metric for quantifying product-level circularity. J Ind Ecol 21(3):545–558
7. Saidani M, Yannou B, Leroy Y et al (2017) How to assess product performance in the circular economy?

Proposed requirements for the design of a circularity measurement framework. Recycling 2(1):6
8. Ghisellini P, Cialani C, Ulgiati S (2016) A review on circular economy: the expected transition to a balanced interplay of environmental and economic systems. J Clean Prod 114:11–32
9. Kirchherr J, Reike D, Hekkert M (2017) Conceptualizing the circular economy: an analysis of 114 definitions. Resour Conserv Recycl 127:221–232
10. Corona B, Shen L, Reike D et al (2019) Towards sustainable development through the circular economy - a review and critical assessment on current circularity metrics. Resour Conserv Recycl 151(104):498
11. Moraga G, Huysveld S, Mathieux F et al (2019) Circular economy indicators: what do they measure? Resour Conserv Recycl 146:452–461
12. Bocken NM, De Pauw I, Bakker C et al (2016) Product design and business model strategies for a circular economy. J Ind Prod Eng 33(5):308–320
13. Kristensen HS, Mosgaard MA (2020) A review of micro level indicators for a circular economy-moving away from the three dimensions of sustainability? J Clean Prod 243(118):531
14. Cong L, Zhao F, Sutherland JW (2019) A design method to improve end-of-use product value recovery for circular economy. J Mech Des 141(4)
15. Favi C, Germani M, Luzi A et al (2017) A design for EoL approach and metrics to favour closed-loop scenarios for products. Int J Sustain Eng 10(3):136–146
16. Zwolinski P, Lopez-Ontiveros MA, Brissaud D (2006) Integrated design of remanufacturable products based on product profiles. J Clean Prod 14(15–16):1333–1345
17. Scheepens A, Vogtlander J, Brezet J (2015) Two life cycle assessment based methods to analyse and design complex circular economy systems. Case: making water tourism more sustainable. J Clean Prod 5:1–12
18. Di Maio F, Rem PC, Baldé K et al (2017) Measuring resource efficiency and circular economy: a market value approach. Resour Conserv Recycl 122:163–171
19. Ellen MacArthur Foundation (2015) Circular indicators: an approach to measuring circularity
20. Figge F, Thorpe AS, Givry P et al (2018) Longevity and circularity as indicators of eco-efficient resource use in the circular economy. Ecol Econ 150:297–306
21. Franklin-Johnson E, Figge F, Canning L (2016) Resource duration as a managerial indicator for circular economy performance. J Clean Prod 133:589–598
22. Vanegas P, Peeters JR, Cattrysse D et al (2018) Ease of disassembly of products to support circular economy strategies. Resour Conserv Recycl 135:323–334
23. Mandolini M, Favi C, Germani M et al (2018) Time-based disassembly method: how to assess the best disassembly sequence and time of target components in complex products. Int J Adv Manuf Technol 95(1):409–430
24. Marconi M, Germani M, Mandolini M et al (2019) Applying data mining technique to disassembly

sequence planning: a method to assess effective disassembly time of industrial products. Int J Prod Res 57(2):599–623

25. Das SK, Yedlarajiah P, Narendra R (2000) An approach for estimating the end-of-life product disassembly effort and cost. Int J Prod Res 38(3):657–673

26. van Loon P, Van Wassenhove LN (2018) Assessing the economic and environmental impact of remanufacturing: a decision support tool for OEM suppliers. Int J Prod Res 56(4):1662–1674

27. Vogtlander JG, Scheepens AE, Bocken NM et al (2017) Combined analyses of costs, market value and eco-costs in circular business models: eco-efficient value creation in remanufacturing. J Remanuf 7(1):1–17

28. van Schaik A, Reuter MA (2016) Recycling indices visualizing the performance of the circular economy. World Met Erzmetall 69:5–20

29. McDonough Braungart Design Chemistry (2013) Cradle to cradle certified product standard version 3.0

30. Di Maio F, Rem PC et al (2015) A robust indicator for promoting circular economy through recycling. J Environ Prot 6(10):1095

31. Mesa J, Esparragoza I, Maury H (2018) Developing a set of sustainability indicators for product families based on the circular economy model. J Clean Prod 196:1429–1442

32. Lee HM, Lu WF, Song B (2014) A framework for assessing product end-of-life performance: reviewing the state of the art and proposing an innovative approach using an end-of-life index. J Clean Prod 66:355–371

33. Alamerew YA, Brissaud D (2019) Circular economy assessment tool for end of life product recovery strategies. J Remanuf 9(3):169–185

34. Niero M, Kalbar PP (2019) Coupling material circularity indicators and life cycle based indicators: a proposal to advance the assessment of circular economy strategies at the product level. Resour Conserv Recycl 140:305–312

35. Periyasamy A, Wiener J, Militky J (2017) Life-cycle assessment of denim. In: Sustainability in denim, pp 83–110

36. Levi Strauss & Co (2015) The Life Cycle of a Jean: Understanding the environmental impact of a pair of Levi's® 501® jeans

37. Lee AWL, Neo ERK, Khoo ZY et al (2021) Life cycle assessment of single-use surgical and embedded filtration layer (EFL) reusable face mask. Resour Conserv Recycl 170(105):580

38. Gross B, Zheng Z, Liu S et al (2020) Spatio-temporal propagation of COVID-19 pandemics. Europhys Lett 131(5):58,003

39. Seidi F, Deng C, Zhong Y et al (2021) Functionalized masks: powerful materials against COVID-19 and future pandemics. Small, 2102453

40. Moreno M, De los Rios C, Rowe Z, et al (2016) A conceptual framework for circular design. Sustainability 8(9):937

41. Blomsma F, Tennant M (2020) Circular economy: preserving materials or products? introducing the resource states framework. Resour Conserv Recycl 156(104):698

42. Tam E, Soulliere K, Sawyer-Beaulieu S (2019) Managing complex products to support the circular economy. Resour Conserv Recycl 145:124–125

43. Aoe T (2007) Eco-efficiency and ecodesign in electrical and electronic products. J Clean Prod 15(15):1406–1414

44. Golinska-Dawson P, Kosacka M, Werner-Lewandowska K (2018) Sustainability indicators system for remanufacturing. In: Sustainability in remanufacturing operations, pp 93–110

45. Jiang Z, Ding Z, Zhang H et al (2019) Data-driven ecological performance evaluation for remanufacturing process. Energy Convers Manage 198:1–12

46. Virtanen M, Manskinen K, Eerola S (2017) Circular material library - An innovative tool to design circular economy. Des J 20:S1611–S1619

Part III
Towards a Circular Future—Consumers for Granular Circularity

If we don't solve this one problem [i.e. moving from a linear to a circular economy], everything else we do, no matter how well-intentioned it is, will be like shifting deck chairs on the Titanic.

—Prof. Dr. Wayne Visser

Professor and Director
Sustainable Transformation Lab, Antwerp Management School
Fellow, Cambridge Institute for Sustainability Leadership

Consumer-Centric Circularity: Conscious Changes in Lifestyle Towards a New Normal

7

Sven Kevin van Langen, Patrizia Ghisellini, Rashmi Anoop Patil and Seeram Ramakrishna

Abstract

In the final chapter, we take a closer look at the role that consumers play in the transition to the CE. Ultimately, consumers must be willing to buy sustainable circular products over those products that have a single lifecycle. The CE can receive great impulses even from a single person or a small group of people as we have shown in the repair café and iFixit case studies. This can cause a ripple effect that can quickly reach millions of people all over the world. It is up to governments to spread awareness amongst their citizens, so they become responsible consumers, and play a role as one of the largest consumers in their own economy as well. We look at the city of Rotterdam for a collection of cases, specifically looking at several consumer initiatives and how the city has been improving its own consumption since governments are big consumers themselves. We end the chapter, and the book, by looking at the circularity gap and revisiting what has been discussed in this book regarding closing this gap.

Sven Kevin van Langen and Patrizia Ghisellini contributed equally to this work.

S. K. van Langen(✉)
UNESCO Chair in Environment, Resources and Sustainable Development (International Ph.D. Programme), Department of Science and Technology, Parthenope University of Naples, 80143 Naples, Italy
e-mail: kevin.vanlangen@studenti.uniparthenope.it

Olympia Electronics, Thessaloniki, Greece

P. Ghisellini(✉)
Department of Engineering, Parthenope University of Naples, 80143 Naples, Italy
e-mail: patrizia.ghisellini@gmail.com

R. A. Patil · S. Ramakrishna(✉)
The Circular Economy Task Force, National University of Singapore, Singapore 117575, Singapore
e-mail: seeram@nus.edu.sg

S. Ramakrishna
Department of Mechanical Engineering, National University of Singapore, Singapore 117575, Singapore

Centre for Nanotechnology and Sustainability (NUSCNS), 2 Engineering Drive 3, Singapore 117576, Singapore

Keywords

Consumer-centric circularity · Consumer engagement · Embracing circularity · Circular lifestyle · Closing circularity gap

7.1 Introduction

This chapter seeks to explore the role of consumers and their influence in the transition towards the CE. The consumers' demand for

R. A. Patil and S. Ramakrishna (eds.), *Circularity Assessment: Macro to Nano*,
https://doi.org/10.1007/978-981-19-9700-6_7

products and materials is an important dimension that captures their role in the transition. At the global level, over the past few decades, this demand for products has increased exponentially contributing to the huge exploitation of natural resources and the rise of global environmental problems such as climate change [1]. The UN, The EU, and national governments are increasing their efforts in addressing the challenges posed by the increasing consumption as part of the transition to the CE [2–4]. The EU, in its CE Action Plan, stresses that it is central to take into account the consumers' perspective and promote their engagement [3]. The UN' 10-Year Framework Programs on Sustainable Consumption and Production (10YFP) adopted in 2012 at the World Summit on Sustainable Development includes an action plan on consumer engagement and a series of targets for promoting responsible consumption under the Sustainable Development 2030 agenda (SDG 12).[1]

This chapter begins by focusing on the 'ripple effect' (by considering the concept positively and constructively) in the current move to a CE and highlighting how a small change in the demand by individual consumers could generate larger and impactful effects in accelerating the adoption of the CE. This is because today's globalized market with millions of consumers gives them the power to enforce change through their individual choices that either support or hamper the transition phase. Then, an overview of initiatives by governments and others to promote circular and sustainable consumption is provided followed by the description of indicators developed to monitor the signs of progress in consumption patterns towards more sustainable models involving circular principles.[2] These could also imply radical changes in people's lifestyles given that CE is a driver towards sustainable development where environmental and social dimensions also play a role in the development path [5]. Further, the major drivers, challenges,

and benefits of a circular lifestyle [6] that enable influencing and steering consumers towards a CE are presented. In this chapter, several case studies are documented. The first case study discusses consumer initiatives such as the Repair Café Foundation, an NGO that provides free repairs with the help of volunteers, and the second, iFixit, a web platform that provides service and repair manuals for free. We also revisit the case of Rotterdam and several initiatives taken in the city by and for consumers. Finally, visions for 2030 and 2050, and the circularity gap are discussed as a conclusion to the chapter and the book.

7.2 Impact of Consumers on Circularity: A Ripple Effect

Historically, firms were normally considered the drivers of paradigm shift through innovation [7, 8]. The impact of consumers and their role as co-developers is a recent field of study in the product and market innovation domain [9]. This study suggests that there are many ways in which consumers can influence product innovation and create markets. It can take one single person to start a movement to change the economy! For example, Greta Thunberg singly tried to influence the Swedish Government to consider and act in mitigating climate change. Her initiative gradually turned into a movement (the ripple effect) as students and activists across Europe joined and compelled their Governments to take measures to achieve carbon reduction and offsetting.

Other examples of the ripple effect related to sustainability are the observation of The Earth Day[3] on April 22nd of every year and The Earth Overshoot Day. The Earth Day was first organized in 1970 as a local environmental movement that has grown into a globally celebrated one, with an intention of becoming more sustainable. The Earth Overshoot Day is that day in each year when resource consumption exceeds the earth's capac-

[1] Issue brief SDG 12: Ensuring sustainable consumption and production patterns https://wedocs.unep.org/bitstream/handle/20.500.11822/25764/SDG12_Brief.pdf.

[2] CE Indicators https://ec.europa.eu/environment/ecoap/indicators/circular-economy-indicators_en.

[3] Earth Day is 50—Kimberlee Hurley https://www.sustainability-times.com/environmental-protection/earth-day-is-50-we-have-our-work-cut-out-for-the-next-50-years/.

ity to regenerate those resources over the year.[4] That means, each day after this, we are creating an ecological deficit by drawing the resource stocks and operating in an 'overshoot' which is unsustainable in the long run. The Earth overshoot day is gaining an increasing amount of media attention each year. It serves as a simple indicator of how human society is increasingly draining the ecological wealth on earth.

The precept of the ripple effect is that a single person or small group, or perhaps a single action, can be the drop that spreads far-reaching ripples throughout societies and the world, echoing into the future. In economic markets, the ripple effect often starts when a consumer notices a problem and seeks to fix it. Three types of logic are identified to changing a market, as listed below.

1. Incumbent legitimator logic, where existing providers (often firms) work with external stakeholders such as consumers;
2. Consumer activist origin, where consumers themselves take the lead and change or start a market;
3. Co-creator scenario, where existing firms and consumers innovate in tandem [10].

All three aforementioned cases can start with one consumer, or perhaps a small group that ends up having a far-reaching effect that spills over well beyond solving their initial problem. In the context of the CE, we are witnessing consumers creating such ripple effects and steering the economy towards circularity. Although the ripple effect in the aforementioned examples seems to have spread quickly in a short time, it generally does take longer. Nonetheless, this can have a profound impact and for that reason, it is important to study trends of consumerism and understand their implications on the economy. With this context, we will study two cases of consumers who started organizations aimed at increasing circularity in the consumer electronics sector through making repairs more accessible, but in very different ways.

7.2.1 Case Study I: Repair Cafés and iFixit

In recent years, the market for electronics repairs has been shrinking in the developed countries. The main reason for this is that electronic products are becoming increasingly complex, making them difficult to be repaired and requiring highly skilled personnel and sophisticated tools and spare parts [11–13]. In this case study, we look at the non-profit Repair Café Foundation and the for-profit iFixit company, two organizations founded by groups of consumers who intend to make repairs easier.

In 2009, Martine Postma set up the first Repair Café in Amsterdam.[5] As a former environmental journalist, she was well aware of how much waste society generates. She realized consumers tend to quickly throw away end-of-life products, to be replaced by something new, even if a simple repair could have extended that product's lifespan. She got inspired by a design exhibit[6] on the benefits of repairs and wanted to help people repair their products. This led to the first Repair Café in which a group of people with repair experience (most of them elderly and retired), came together in a conference room and helped neighbors. They quickly moved to a nearby community center and received increased attention from (social) media.

In 2011, Ms. Postma started the non-profit Repair Café Foundation that provided support in setting up new Repair Cafés. It started with new Repair Cafés in the Netherlands (Fig. 7.1) but quickly spread throughout Europe and other parts of the world. At the time of writing (2021), there are over 2000 Repair Cafés in the world, ~90% of which are in Europe.[7] Repair Cafés are present in almost every municipality in the Netherlands[8] The Repair Café Foundation has also started a repair monitor program to track repairs performed

---

[4] About Earth Overshoot Day https://www.overshootday.org/about-earth-overshoot-day/.

[5] An Effort to Bury a Throwaway Culture One Repair at a Time https://www.nytimes.com/2012/05/09/world/europe/amsterdam-tries-to-change-culture-with-repair-cafes.html.

[6] Platform21 https://www.platform21.nl/page/133/en.html.

[7] Repair Café https://www.repaircafe.org/en/visit/.

[8] Ten years of Repair Café https://nos.nl/artikel/2306621-tien-jaar-repair-cafe-al-ruim-een-miljoen-producten-gered-van-afvalberg.

Fig. 7.1 An overview of
Repair Café's in the
Netherlands from 2019, 8
years after the founding of
the Repair Café
Foundation. Copyright
OpenStreetMap
contributors
(see Footnote 8)

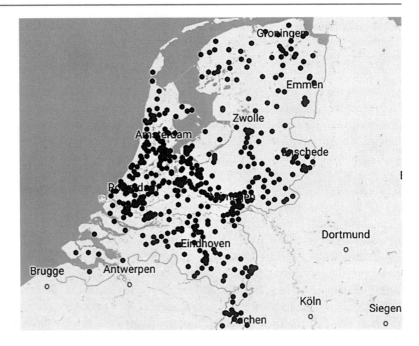

in their cafés and gain quantitative data on product repairs.

Besides providing just repairs, Repair Cafés also serve social and educational roles. This is in line with other such 'open workshop' initiatives, which have a great potential to enable a granular CE beyond recycling [14, 15]. These alternative methods for reducing waste, such as repair, reuse, remanufacturing, and regeneration, originate primarily from small organizations that are often non-profit in nature. According to the studies, small organizations, such as the Repair Cafés, play an important role in raising consumer awareness on the CE [14, 15]. The high growth of Repair Cafés shows that there is still much potential in embracing the CE.

iFixit also provides a platform to facilitate the repair of consumer electronics. They provide a digital platform with standardized service/repair manuals and complete product teardowns provided by their team or by the community [16]. iFixit[9] was started by two consumers who tried to fix their laptops but could not find instructions or spare parts. The founders, Luke and Kyle, wanted to provide easy-to-use repair manuals for free and

also provide an online web store for spare parts and tools. Since its start in 2003, iFixit has become a platform with over 70,000 open access manuals for more than 30,000 devices. Together with instructional videos on YouTube, iFixit is one of the leading platforms for providing information on repairing consumer electronics.

iFixit also serves some social functions, such as in education.[10] It is increasingly being used as an educational tool, integrated into educational courses [17–19]. It serves as an example of technical documentation and can be used interactively by instructors. This introduces students to circular practices by teaching them in-demand skills. iFixit has also become a database for qualitative and quantitative research as they gather narratives on successful and unsuccessful repairs. This has been used to analyze how consumers typically detect faults with certain products [20], or to learn statistical data about common faults and repairs that consumers face regarding their products. The geographical distribution of the platform's users (as of 2020) is shown in Fig. 7.2 [21, 22].

---

[9] iFixit https://www.ifixit.com/Info/background.

[10] iFixit Technical Writing Project https://edu.ifixit.com.

Fig. 7.2 A study by J. Lepawsky mapped the distribution of iFixit's active community members [22]. Copyright Mapbox and OpenStreetMap contributors

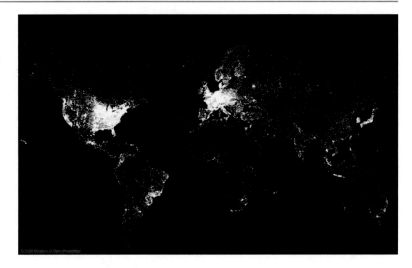

7.3 Embracing Circularity: The Role of Governments

When it comes to governments and their relation to consumerism, there are two major roles that a government can play [23]. First, there is the government itself as a consumer. Governments are usually one of the biggest spenders within any economy as they indulge in public procurements (purchase of goods and services using public funds). Therefore, a government must align its procurements to its targets in achieving a CE. Public procurement not only creates a direct demand for circular products/services, but also reinforces ties between circular businesses, improves their market position, and can impact larger parts of a supply chain [24]. Other consumers such as private businesses and individuals might also follow their government by investing in circular products [25].

Governments can also play a vital role in maneuvering consumer behavior through various initiatives such as imparting awareness on the CE, subsidizing circular products/services, amending regulations that form a barrier for unsustainable consumer activities [26]. Subsidies have a strong effect on the buying behavior of consumers [25]. Governments and their agencies can influence consumers by conveying information on the impact and availability of circular products directly and indirectly [27]. They can provide information platforms to guide consumers in

creating a positive impact using their buying power or facilitate the creation of new labels or certifications that instill trust in products from recycled materials and remanufactured products. Governments can also exercise their authority to introduce new regulations/legislation or amend the existing ones that help stimulate the consumption of new circular products and services. As mentioned earlier, governments can monitor the societal behavior regarding CE practices using indicators. For instance, the EU actively monitors consumer activities through the following indicators[11]:

- **Percentage of citizens that have chosen alternatives to buy new products**: The alternatives considered are the purchasing of a remanufactured product, the leasing or renting of a product instead of buying it, used sharing schemes (e.g. car or bike-sharing);
- **Coverage of the CE topic in electronic mass media, number of articles published**: This indicator aims to track the interest in electronic mass media of the CE and its popularity across citizens.

Furthermore, classical economic indicators are also proposed in the same set of the above indicators related to:

---

[11] Societal behaviours https://ec.europa.eu/environment/ecoap/indicators/societal-behaviours_en.

- **The annual turnover in the repair of computers and personal goods**: The monitoring of repair is strategic in the move to CE since it extends the service life of products contributing to reducing the consumption of natural resources and the production of waste and the associated negative environmental impacts. The benefits are also socio-economic as the repair services create new economic and job opportunities. These benefits are tracked using the next indicator;

- **The number of enterprises and employment in the repair of computers and personal and households goods**: This indicator monitors across the EU countries the move to more circular business models and is then useful to evaluate the structural changes in the economy of the member states and the whole EU.

7.3.1 Case Study II: Circular Rotterdam—Consumer Initiatives

In Chap. 2, we discussed the case of Rotterdam's approach to becoming fully circular by 2050. we now build on that case study by looking at several of the initiatives aimed at consumers that are supported by the city.[12] Much of the city's efforts are aimed at encouraging bottom-up activities originating from entrepreneurs and existing businesses, many of which are focused on consumer issues. From the city's web platform on their circular projects,[13] we specifically filtered for 'citizen initiatives.[14] This gives us two cases to study in this section more in-depth to see how citizens helped the transition to the CE. Furthermore, we will look at several instances where the municipality of Rotterdam has committed to circular actions in its role as a consumer. Finally, we

look at a way that the city government has tried to influence consumers.

The first case is called Wijkea, an initiative by volunteers to reallocate used furniture to people on welfare support/low income.[15] It is specific to the Kralingen neighbourhood of Rotterdam and provides free pick-ups of old furniture and redistributes it to people that need it most, willing to accept used furniture with some wear. Furniture can be both given for free permanently or it can be temporarily loaned. Volunteers also help in restoring damaged furniture or maintenance (e.g., with a paint job). These restoration projects are just to teach skills to both people with long-term unemployment and interested volunteers, adding a socio-economic dimension to the project.

The second case is called Urban Mining with Credits, a start-up from Rotterdam that offers a form of crowdfunding where you gain shares, in the form of credits, in exchange for your electronic or polymer waste,[16] this project is initiated in the nearby city of The Hague as well.[17] Every kilo of e-waste or plastic equals one credit. The collected waste is disassembled and sold for reuse, upcycling, or recycling. After 30 months, any operational costs are deducted from the returns on the waste sales, the profits are redistributed to all people with credits. As with the Wijkea project, the labor for Urban Mining with Credits is mostly performed by people trying to build employable skills. Some of the targeted buyers of disassembled household waste are electronics repair stores that need spare parts and people that make 3D printer filament from shredded plastics and other polymers.

One area where Rotterdam tries to implement circular procurement, acting as a consumer, is road construction. There are two projects in this area. The first is the re-use of old concrete of decommissioned roads that is reused in the construction

---

[12] From trash to treasure: Rotterdam Circularity Programme 201–2023 https://rotterdamcirculair.nl/wp-content/uploads/2019/05/Rotterdam_Circularity_Programme_2019-2023.pdf.

[13] Rotterdam Circularity Initiatives Archive https://rotterdamcirculair.nl/initiatieven/.

[14] Citizen initiative is a translation of 'burgerinitiatief', one of the pre-set filtering options on rotterdamcirculair.nl portal.

---

[15] https://vrijwilligerswinkel.nl/servicediensten/wijkea/.

[16] Urban mining with credits https://citylab010.nl/initiatieven/urbanminingwithcredits.

[17] Urban Mining with Credits—ImpactCity https://impactcity.nl/innovator/urban-mining-with-credits/.

of new bicycle lanes.[18] The new lanes are built, at all major layers, from recycled concrete mined from Rotterdam itself that is rebound using a bio-based resource that is a byproduct of the paper industry. Though in a way it is downcycling of the concrete, as bicycle roads have to deal with less weight than regular roads, it can still be fully utilized. In another road construction project, the city uses rubber that is recycled from old car tires to build more silent roads in residential zones.[19] The use of recycled rubber for this purpose has proven to build durable roads that are 10 dB more silent than non-rubberized roads. Outside of road construction, the city is experimenting with greener alternatives to classical cement-based concrete. In one project, the cement in the concrete mixture was replaced by local household waste and used for building public benches.[20] In its production, a lot less CO_2 is emitted than with regular concrete, an additional benefit is that cement is normally hard to recycle[21] while the household waste is in a second cycle already in this way. The initial research was funded by the municipality of Rotterdam and performed by the nearby Delft University, the city also is the first consumer of this new type of circular concrete for public benches and is studying its use for the construction of bridges.

Finally, by providing a web platform (see Footnote 13), the city is influencing consumers actively. Rotterdam's city government has made a website where they provide news, information, and a collection of CE-related initiatives with the purpose to engage its citizens in the city's transformation. They offer a monthly newsletter and are active on major social media platforms with accounts highlighting the developments in the transition to a CE. Their site provides information on how to be more circular as a consumer,

and which nearby organizations might help with that. Providing several filtering options to find many different types of initiatives. Furthermore, the site provides information on who to contact if you have an idea you want to turn into a new initiative and where to get subsidies.

7.4   Circular Lifestyle: Challenges and Benefits

Consumers both in the EU and at the global level are becoming increasingly concerned about the environment and its problems[22] and are also aware of their consumption habits having negative effects on the environment.[23] In the EU, consumers also show good awareness and attitude of the CE concept even if not uniform across the member states. Such attitude has increased during the Covid-19 pandemic since it can be considered an effect of the conflictual relationship between humans and nature. Consumers think it is important to reduce the use of resources, reuse products at the end-of-life and increase the lifetime of products.[24] Consumers are also aware of the positive impact of purchasing environmentally-friendly products on the environment and society (see Footnote 24). Changing consumption habits is one of the most effective ways of taking action to tackle environmental issues. In this regard, the concept of a circular lifestyle has also emerged. A circular lifestyle is one where consumer purchases and acts in such a way that a minimum amount of resources is lost from an economic system. Proponents and critiques of CE alike claim that CE on its own cannot succeed without lifestyle changes. We already discussed consumer behavior towards preferring circular products, but we must make more radical lifestyle changes to have the CE succeed [5]. Typically, this is done by following an xRs typology [28], the exact amount of 'R's often

---

[18]   Recycled asphalt in Rotterdam https://rotterdamcirculair.nl/initiatieven/blue-city-4/.

[19]   Roadways in Rotterdam from re cycled car tires https://rotterdamcirculair.nl/initiatieven/rubberpave/.

[20]   Rotterdam builds bank of 'green concrete' together with TU DELFT https://rotterdamcirculair.nl/nieuws/rotterdam-bouwt-samen-met-tu-delft-bank-van-groen-beton/.

[21]   Concrete and Circular: (How) Do They Go Together? http://rotterdamcirculair.nl/nieuws/beton-en-circulair-hoe-gaat-dat-samen/.

[22]   Meet the 2020 consumers driving change https://www.ibm.com/downloads/cas/EXK4XKX8.

[23]   Attitudes of Europeans towards the environment https://europa.eu/eurobarometer/surveys/detail/2257.

[24]   https://www.art-er.it/2020/12/certificazioni-ambientali-stato-della-rte-e-prospettive-evolutive-disponibili-i-materiali-del-webinar/.

varying from advice to advice. Typical advice to achieve a more circular lifestyle is listed below.

1. **Rethink** the first R, entails that consumers rethink the vision of natural resources as available in limited supply and make less impact on them in everyday life.

2. **Refuse** is the practice of not buying products from companies that harm the environment. This can come from companies that have had negative publicity, but also by looking at how companies offer the products (e.g., packed in more plastic than necessary).

3. **Reduce** is the practice of using fewer resources. e.g., by replacing single-use plastics with durable replacements or saving leftovers.

4. **Repurpose** is the practice when one uses a product for a new purpose. This can happen for several reasons, perhaps the original use was not required anymore, or the product degraded to a point that it cannot be used for its initial function anymore.

5. **Reuse** is the practice of using products that have a previous owner. Often people lose the need for a product while it is still perfectly functional and buying such products on secondary markets can prevent an entire product lifecycle.

6. **Repair** is to fix a broken product. One can fix their product and keep using it, or have it fixed, but one can also buy broken products and repair them by themselves.

7. **Recycling** a product implies that the product is brought down to its base materials for use in the production of a new product. While this was typically hard to implement as a consumer, in some traditional crafts (e.g., carpentry) it is a viable option. Modern developments in 3D printing even make it possible to upcycle single-use plastics into durable products.

8. **Rot** also known as composting, is similar to recycling but specifically for food waste and/or gardening waste to replenish soils with fresh nutrients.

There is increasing evidence suggesting that circular practices alone will not decrease the consumption of primary materials [29]. These practices might increase the consumption of primary materials and this is termed as the CE Rebound Effect. By increasing the value at the end of the (first) lifecycle of a product through CE practices (as the 'waste' can still be used), one can also increase the value further upstream in the value chain and thus increase the supply of primary materials.

One of the proposed solutions to reduce waste is to reduce overall consumption. This is supported by the degrowth movement [5, 30], whereas most policies aimed at achieving a CE, including those of the EU, assuming that a CE will lead to a decoupling of economic growth from resource use [31]. Historically, there has been a strong correlation between the number of primary resources used in an economy and the size of that economy. The CE is often supported to lower that correlation. It was at first expected this would also lead to fewer primary materials being used over time, but now, there is evidence to show this is not necessarily true. The rebound effect might see an increased level of the economy per resource unit used, and not lower the absolute resource use [29]. Supporters of the degrowth movement emphasize the use of raw materials is often harmful to the environment and thus to protect it, societies must be willing to reduce economic growth to remain sustainable. While this theory started at the fringe, it has recently begun to attract some mainstream attention and is now, even studied and considered by the EU and the European Environment Agency[25] as well as by the UN [32]. A study by the EU's Joint Research Centre found that the academic paradigm around degrowth is still evolving, with the latest developments focusing on how to measure its claimed benefits [33]. The Joint Research Centre states that more research should be done into measuring the benefits of degrowth, suggesting tests based on input-output modeling, material flow analysis, life-cycle analysis, or social surveys with a special focus on non-market value creation.

---

[25] https://www.eea.europa.eu/themes/sustainability-transitions/drivers-of-change/growth-without-economic-growth.

Degrowth can be achieved in several ways.[26] One can reduce the consumption by end-users, but another approach is by shrinking the supply chain. One way to shrink the supply chain is through urban farming. A concept that, while not new, has recently witnessed an increased adoption speed, more so in developed countries [34]. Urban farming is a form of urban regeneration where citizens grow their food communally in the city or privately at home. With modern advancements like hydroponics, this has become increasingly accessible. While seeds or saplings are often still procured from classic farms, much of the growing is done near the place of consumption and without any heavy machinery, drastically reducing the economic size of the food supply chain and the amount of (primary) resources involved. In the US, urban farming is practiced mainly by younger and more educated households and is rarely enabling the involved households to manage their consumption completely on their produce. Furthermore, the number of people (involved in urban farming) that abandon the practice is quite high, indicating that urban farming may not be a sustainable practice yet and needs to mature more in terms of implementation [35]. In Italy, urban farming has a long tradition, often being employed to alleviate poverty, and is being supported by local governments [36, 37]. It found its origin in the second world war when the people of Italian cities such as Bologna, Rome, Milan, and Naples faced a shortage of food [37, 38]. The interest in urban farming in Italy continued after the second world war. Moreover, the European debt crisis caused a renewed interest in urban farming in Naples and the rest of Italy, as part of sustainable development initiatives and is also implemented by people of older age [36]. Currently, Bologna has one of the biggest projects of urban gardening in Italy, comprising 47 hectares of municipality land. Overall, there are more than 2,750 urban vegetable gardens [37]. Noted developments have taken place in urban food forestry, which, unlike urban farming, is more often communally implemented [38, 39]. The use of land for urban gardens contributes to

putting in circulation again urban areas that could otherwise end up abandoned due to degradation or subject to possible building speculations. In this way, urban gardens are contributing to change and redefine the landscape of a city according to the principles of the CE [37].

7.5 Closing the Circularity Gap: Vision and Road Ahead

As more governments pledge climate targets for 2030 and 2050-2060, with many of the major economies in the world aiming to be completely climate neutral by 2050–2060, a CE will play an increasingly important role in achieving these targets. The savings in greenhouse gas (GHG) emissions as a result of circular practices will greatly reduce the investments required for the energy transition. To achieve the targets of the Paris climate agreement, a lot of the GHG emission reduction should be achieved by 2030, and to do so, transitioning to a CE can be a powerful driver, with a strong return on investment. While not every government is aiming to be fully circular at the same time as they become climate-neutral, however, transitioning to the CE has become a part of their strategy. Nonetheless, governments that don't aim to become completely circular soon, do still recognize that they should take significant steps in closing the major circularity gap.

It is expected that governments (both national and local) will play a leading role in the societal transitions to a CE. Therefore, it is important that governments properly measure and monitor the transition, to effectively support the businesses and citizens they represent. A good first step would be to understand the material flows in and out of their economy at the national level. From here, it is crucial to, pick the low-hanging fruits so the transition can start swiftly and have a strong impact on emissions and improving ecologies, and then, assessments can be made on how to tackle the more difficult issues of the transition. It will be important to iterate the assessment process to find the optimum interventions that will move society further towards circularity. If enough cities are willing to pledge maximum circularity

---

[26] What is Degrowth? https://www.degrowth.info/en.

by 2050, as Rotterdam is doing, this can get a snowball rolling, and best practices can be shared to eventually facilitate all cities to reach the target of resource neutrality.

As discussed in Chap. 1, the global economy is only 8.6% circular as of 2020 [40]. We are heavily dependent on virgin resources and have a relatively low capacity to process end-of-life products for a return to reuse in the economy. To support a transition to a CE we need to better measure our efforts to make this happen and for that, we need quantitative indicators. This needs to happen at the macro, meso, micro, and nano levels (Fig. 1.1). To this purpose, ISO 59020 is currently in development.

It was evident from Chap. 2 that the evaluation and reporting of ESG performance of businesses overlaps with circularity assessment at the micro-level (in the systemic hierarchy), as most of the measured indicators for environmental and social aspects are common to both. It is, therefore, important to take note of the loopholes in the ESG reporting standards and regulations for developing a robust and universally applicable circularity assessment standard. On the other hand, circularity assessment should become a part of sustainability reporting to tap into the financial opportunities for circular projects, currently available for sustainable development.

At the macro level, as shown in Chap. 3, we find that MFA, Input-Output Analysis, and LCA as popular methods of assessments, with the best results achieved when several methods are combined. Much research is especially focused at the city level, for which several initiatives have been launched. Often indicators used by circular cities relate to the (relative) amount of materials that are recycled, the CO_2 emissions saved, the amount of waste that is avoided by initiatives, and the number of new jobs generated. Table 3.1 presents a large overview of indicators used by cities. On a larger scale, like the nation or for the EU, indicators tend to be more focused on per capita and relative figures over absolute amounts, as shown in Table 3.2.

At the meso level, where economic sectors are discussed, we typically highlight initiatives related to EIPs, as shown in Chap. 4. With EIPs, a synergy is sought between companies in different parts of supply chains, to minimize the externalities of the industrial park as a whole. First of all, a way to identify EIPs is presented in Chap. 3. We found several assessment frameworks, such as the UNIDO requirements showcased in Table 4.1 and the National EIPs Evaluation Standard System of China that is presented in Table 4.2. The UNIDO framework is more of a checklist for best practices for EIPs while the Chinese evaluation system is more focused on measuring the performance of EIPs.

At the micro-level, we look at individual companies as shown in Chap. 5. While most studies consider products and components as part of the micro-level, we follow the ISO/TC323 explanatory note and put products and components as a separate economic hierarchy 'nano level' in Chap. 5. For businesses, there are distinct indicators in the areas of operations (i.e., resource efficiency, eco-design, and green projects), waste management (i.e., wastewater, emissions, solid waste, recycling, remanufacturing, and end-of-life management), and supply-chain management. This is because when companies belong to diverse economic sectors, the indicators relevant and most impactful for each of these companies differ highly—a primary reason as to why, often, companies develop their own set of indicators.

At the nano-level, individual commercial products are assessed for their environmental impacts majorly using the LCA as shown in Chap. 6. Although the LCA is not truly a circular assessment, it is vastly applied to a variety of products and is closest to a circularity assessment. The indicators measurements based on the life cycle of a particular product are categorized into standard impact categories (such as contributing to climate change, water bodies eutrophication, and toxicity caused in the biota) useful for assessing the product. However, this approach does not cover the economic dimension and the circularity at the grass-root level. Therefore, revisiting the design of existing products from the perspective of material efficiency criteria becomes critical in achieving circularity.

Much will also depend on consumers and if they are willing to change their lifestyles. If they continue in their current path of preferring new products and low prices, it is likely society will remain mostly in a recycling economy and will not adopt a more granular approach that includes reducing consumption and enhanced product reuse. Consumers also hold the power to push firms and governments to make an effort in transitioning to the CE. People must have the right awareness so they can collectively steer society in the way they deem best. This can even include an economic viewpoint where (short-term) growth is no longer a key driver in policy, but more focus might be put on sustainability instead, with a stronger focus on environmental and social dimensions.

Most indicators used for the CE at the micro or meso level focus on the production and reverse supply chain of goods and waste, with only macro indicators capturing consumption patterns. There is little detailed data available on the actual consumption and real durability of goods [41]. When products reach consumers, it cannot be assumed they are being actively used [42]. Furthermore, after initial consumption, most of the second-use transactions are within the informal economy which makes tracking of reuse difficult and challenging [22]. Also in the recycling industry, the large informal sector active in this area can make it challenging to properly measure how consumers treat their end-of-life goods [43, 44]. To gain a better idea of how circular our society is and to further the transition to a CE, it is crucial to trace products and materials through all life cycle stages, including the use phase with consumers.

References

1. Camacho-Otero J, Boks C, Pettersen IN (2018) Consumption in the circular economy: a literature review. Sustainability 10(8):2758
2. United Nations (2019) The sustainable development goals report
3. The European Union (2020) Circular economy action plan: for a cleaner and more competitive Europe
4. Shao J (2019) Sustainable consumption in china: new trends and research interests. Bus Strateg Environ 28(8):1507–1517
5. Hobson K, Lynch N (2016) Diversifying and degrowing the circular economy: radical social transformation in a resource-scarce world. Futures 82:15–25
6. Pisoni A, Michelini L, Martignoni G (2018) Frugal approach to innovation: state of the art and future perspectives. J Clean Prod 171:107–126
7. Schumpeter JA (1935) The analysis of economic change. Rev Econ Stat 17(4):2–10
8. Schumpeter JA et al (1939) Business cycles, vol 1. McGraw-Hill New York
9. Gemser G, Perks H (2015) Co-creation with customers: an evolving innovation research field. J Prod Innov Manag 32(5):660–665
10. Branstad A, Solem BA (2020) Emerging theories of consumer-driven market innovation, adoption, and diffusion: a selective review of consumer-oriented studies. J Bus Res 116:561–571
11. Karagiannidis A, Perkoulidis G, Papadopoulos A et al (2005) Characteristics of wastes from electric and electronic equipment in Greece: results of a field survey. Waste Manag & Res 23(4):381–388
12. McCollough J (2009) Factors impacting the demand for repair services of household products: the disappearing repair trades and the throwaway society. Int J Consum Stud 33(6):619–626
13. Ylä-Mella J, Keiski RL, Pongrácz E (2015) Electronic waste recovery in Finland: Consumers' perceptions towards recycling and re-use of mobile phones. Waste Manage 45:374–384
14. Ghisellini P, Ulgiati S (2020) Circular economy transition in Italy. Achievements, perspectives and constraints. J Clean Prod 243:1–18
15. Rabadjieva M, Butzin A (2020) Emergence and diffusion of social innovation through practice fields. Eur Plan Stud 28(5):925–940
16. Wiens K (2011) Dozuki brings technical manuals into the 21st century: a breakthrough in the world of documentation. IEEE Cons Electr Mag 1(1):39–42
17. Getto G, Franklin N, Ruszkiewicz S (2014) Networked rhetoric: ifixit and the social impact of knowledge work. Techn Comm 61(3):185–201
18. Getto G, Labriola JT (2016) IFixit myself: user-generated content strategy in "the free repair guide for everything". IEEE Trans Prof Commun 59(1):37–55
19. McCrigler B, Rippens M (2016) Industry innovation and classroom constraints: infusing real-world ux into the university classroom via iFixit's technical writing project. Int J Sociotechnol Knowl Devel (IJSKD) 8(3):15–28
20. Arcos BP, Bakker C, Flipsen B et al (2020) Practices of fault diagnosis in household appliances: insights for design. J Clean Prod 265:1–11
21. Mashhadi AR, Esmaeilian B, Cade W et al (2016) Mining consumer experiences of repairing electronics: product design insights and business lessons learned. J Clean Prod 137:716–727
22. Lepawsky J (2020) Planet of fixers? Mapping the middle grounds of independent and do-it-yourself information and communication technology maintenance and repair. Geo: Geograph Environ 7(1):1–17

23. Khan S, Haleem A (2021) Investigation of circular economy practices in the context of emerging economies: a CoCoSo approach. Int J Sustain Eng 1–11

24. Rainville A (2021) Stimulating a more circular economy through public procurement: roles and dynamics of intermediation. Res Policy 50(4):104,193

25. Zhou Q, Yuen KF (2020) Analyzing the effect of government subsidy on the development of the remanufacturing industry. Int J Environ Res Public Health 17(10):3550

26. van Langen SK, Passaro R (2021) The Dutch green deals policy: applicability to circular economy policies (Under Review)

27. Pisitsankkhakarn R, Vassanadumrongdee S (2020) Enhancing purchase intention in circular economy: an empirical evidence of remanufactured automotive product in Thailand. Resour Conserv Recycl 156(104):702

28. Reike D, Vermeulen WJ, Witjes S (2018) The circular economy: new or refurbished as CE 3.0?-exploring controversies in the conceptualization of the circular economy through a focus on history and resource value retention options. Resour Conserv Recycl 135:246–264

29. Zink T, Geyer R (2017) Circular economy rebound. J Ind Ecol 21(3):593–602

30. Friant MC, Vermeulen WJ, Salomone R (2020) A typology of circular economy discourses: navigating the diverse visions of a contested paradigm. Resour Conserv Recycl 161(104):917

31. Pineiro-Villaverde G, García-Álvarez MT (2020) Sustainable consumption and production: exploring the links with resources productivity in the EU-28. Sustainability 12(21):8760

32. United Nations Development Programme (2020) The next frontier: human development and the Anthropocene

33. Weiss M, Cattaneo C (2017) Degrowth-taking stock and reviewing an emerging academic paradigm. Ecol Econ 137:220–230

34. Säumel I, Reddy SE, Wachtel T (2019) Edible City solutions-One step further to foster social resilience through enhanced socio-cultural ecosystem services in cities. Sustainability 11(4):972

35. Chenarides L, Grebitus C, Lusk JL et al (2021) Who practices urban agriculture? An empirical analysis of participation before and during the COVID-19 pandemic. Agribusiness 37(1):142–159

36. Rigillo M, Majello MV (2014) Opportunities for urban farming: the case study of san martino hill in naples, Italy. WIT Trans Ecol Environ 191:1671–1683

37. Ghisellini, P, Oliveira, M, Santagata, R, Ulgiati, S (2019) Dossier on the environmental and economic value of investment projects on urban forests and green infrastructures

38. Rusciano V, Civero G, Scarpato D (2017) Urban gardening as a new frontier of wellness: case studies from the city of Naples. Int J Sustain Econ Soc Cult Context 13:39–49

39. Brito VV, Borelli S (2020) Urban food forestry and its role to increase food security: a Brazilian overview and its potentialities. Urban Forestry & Urban Greening, p 126835

40. Haigh L, Wit Mde, Daniels C, Colloricchio A., d Hoogzaad J (2021) The circularity gap report 2021 by circle economy

41. Sánchez-Ortiz J, Rodríguez-Cornejo V, Río-Sánchez D et al (2020) Indicators to measure efficiency in circular economies. Sustainability 12(11):1–15

42. Bovea MD, Ibáñez-Forés V, Pérez-Belis V et al (2018) A survey on consumers' attitude towards storing and end of life strategies of small information and communication technology devices in Spain. Waste Manage 71:589–602

43. Ongondo FO, Williams ID, Cherrett TJ (2011) How are WEEE doing? a global review of the management of electrical and electronic wastes. Waste Manage 31(4):714–730

44. Ismail H, Hanafiah MM (2019) An overview of LCA application in WEEE management: current practices, progress and challenges. J Clean Prod 232:79–93

Index

Printed in the United States
by Baker & Taylor Publisher Services